Safety and Reliability in Cooperating Unmanned Aerial Systems

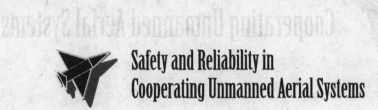

Safety and Reliability in Cooperating Unmanned Aerial Systems

Camille Alain Rabbath

Nicolas Léchevin

Defence R&D Canada - Valcartier

World Scientific

NEW JERSEY • LONDON • SINGAPORE • BEIJING • SHANGHAI • HONG KONG • TAIPEI • CHENNAI

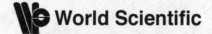

Published by

World Scientific Publishing Co. Pte. Ltd.
5 Toh Tuck Link, Singapore 596224
USA office: 27 Warren Street, Suite 401-402, Hackensack, NJ 07601
UK office: 57 Shelton Street, Covent Garden, London WC2H 9HE

British Library Cataloguing-in-Publication Data
A catalogue record for this book is available from the British Library.

SAFETY AND RELIABILITY IN COOPERATING UNMANNED AERIAL SYSTEMS

ISBN-13 978-981-283-699-1
ISBN-10 981-283-699-3

Printed in Singapore.

Preface

The vision is clear: several unmanned aerial vehicles collaborate and coordinate their flight and actions to achieve a mission, while human operators, barely involved, monitor the progress of the vehicles. This vision is not yet a reality. Before multiple unmanned aerial vehicles are deployed in a coordinated fashion, novel systems must be devised. Among those, systems that ensure safe and reliable operations. Currently, a great many researchers are deploying every effort to design more effective multi-vehicle control concepts and algorithms. Furthermore, there exists a vast body of knowledge in fault-tolerant control, and in fault detection and fault recovery techniques for the individual aerial platform. Yet, very little has been said to date about how to perform reliable and safe autonomous multi-vehicle operations. Indeed, ensuring mission success despite off-nominal, or degraded, operations of mission-critical vehicle components is an open problem which has drawn attention only recently. Despite fault-tolerant control software and hardware embedded onboard air vehicles, overall fleet performance may still be degraded after the occurrence of anomalous events, such as systems malfunctions, damage and failures. As far as we are aware, this book is the first of its kind in presenting a set of basic principles and algorithms for the analysis and design of health management systems for cooperating unmanned aerial vehicles. Such systems rely upon monitoring and fault adaptation schemes. Cooperative health management systems seek to provide adaptation to the presence of faults, from a team perspective, by capitalizing on the availability of interconnected computing, sensing, and actuation resources. There is currently little literature on the safety and reliability for cooperating unmanned aerial systems, although the topic of cooperation for effective fleet monitoring and fault-adaptation purposes is emerging.

This monograph is the culmination of several years of research, and as such is biased with previous results obtained by the authors. We have our own view on the problem of health management, and have addressed a limited number of scenarios. This monograph presents the concepts in the form of theorems, lemmas, propositions, and step-by-step procedures. The health management concepts are illustrated by means of simple examples and numerical simulations of practical UAS operations. Cases of tight formation control and coordinated rendezvous for a network of formations are addressed in this book. Therefore, researchers, academics, graduate students and aerospace engineers, we hope, will appreciate the content.

We wish to thank Defence R&D Canada, and in particular Dr. Alexandre Jouan for his support of this initiative. The first author acknowledges the support of the Natural Sciences and Engineering Research Council of Canada (NSERC). The second author gratefully acknowledges the support of NSERC and the Department of National Defence of Canada in the form of a Visiting Fellowship. We would like to thank Quanser Inc. for providing experimental data to support the modeling of the ALTAV, in particular Drs Jacob Apkarian and Ernest Earon who have shown constant support of our ideas. We have indeed learned quite a lot from their vast knowledge of real-time control systems. We also had the pleasure of collaborating with several academic researchers in the areas of unmanned systems control and fault tolerance. In particular, we would like to thank Professor Youmin Zhang of Concordia University for the many discussions we had on the area of individual vehicle fault-tolerant control, thus improving our understanding of the issues and challenges in such field. We would like to thank Dr. Antonios Tsourdos and his team at Cranfield University (UK). We have had the honor of collaborating with Antonios for several years, which helped us learn more effective techniques of cooperative control, path planning, and guidance. The generous advices of Mr. Jean Bélanger, Dr. Dany Dionne and Professor Pierre Sicard of University of Québec at Trois-Rivières are also gratefully acknowledged. This book was written over a period of one year after working hours and during weekends. Hence, we would like to express our most sincere gratitude and warmful thanks to our families and friends for their support during this intense period of our lives.

C.A. Rabbath and N. Léchevin

Contents

Chapter 1

Introduction

To date, unmanned aerial vehicles, or UAVs, have been operated in real missions with various levels of autonomy [1, 2]. In the future, unmanned and joint manned-unmanned missions are likely to include cooperative sensor networks for search, rescue, and monitoring; collaborative indoor/outdoor surveillance and protection using small, miniature or micro UAVs [3, 4]; as well as cooperating networked unmanned combat aerial vehicles (UCAVs) and weapons for engaging mobile targets in adversarial environments. Furthermore, UAV applications are expected to include firefighting, some level of policing, first-responder support in case of natural disasters, remote sensing, scientific research, and geographical surveying, to name a few. It is commonly acknowledged that the development of UAVs has been partly motivated by the desire to accomplish missions that are too "dull, dirty, or dangerous" for humans. However, there are still some challenging barriers to overcome before the futuristic vision of multiple UAVs, UCAVs and weapons operating cooperatively with other manned vehicles can be realized.

Over the past few years, there has been significant interest in the design of systems that use multiple autonomous agents to cooperatively execute a mission [5–8]. One of the scientific and technological challenges of multi-vehicle control is ensuring efficiency and safety in a context in which the conditions of the vehicles, network and environment are changing, and are potentially abnormal. Under adverse conditions the capabilities of the vehicles may be reduced, compromising mission success and risking the safety of nearby civilian populations.

This book provides basic principles and algorithms for the design and the analysis of health management systems for missions involving cooperating unmanned systems, with the objective of addressing the realistic contingencies encountered in complex or hostile environments. The theory is complemented by case studies and examples of applications featuring small-scale unmanned vehicles, with emphasis on the modeling of realistic dynamics, implementation of algorithms and systems integration. Motivating this book is the fact that overall fleet performance can be degraded by anomalous events even when fault-tolerant control software and redundant (duplicate) hardware have been installed in air vehicles to increase reliability. For example, when severe body damage or actuator faults occur, a large difference between post-fault and pre-fault system dynamics may result in a significant reduction of control authority. The faulty vehicle is then no longer capable of performing its assigned task with the expected level of efficiency, and its role in the mission may need to be re-planned. Designed to enable teams to adapt to degraded operating conditions, cooperative health management (CHM) systems capitalize on the availability of various interconnected resources and on the sharing of key information among the networked entities with minimal involvement of the operating crew.

1.1 Unmanned Aerial Systems

There are several flight-critical components and systems for the UAVs: actuators, control surfaces, engines, sensors, flight computers, and communication devices. Together the platform with its systems and components form an unmanned aircraft system (UAS) [2]. Fig. 1.1 is a conceptual schematic of a typical UAS control system. The platform represents the vehicle body or UAV. The actuators usually consist of motors that drive control surfaces (ailerons, elevators, rudders, fins, canards), which in turn alter the aerodynamic characteristics of the platform. Servomotors are typically used with commercial-off-the-shelf (COTS) small-scale vehicles and radio-controlled aircraft. The actuation block in Fig. 1.1 can also include the propulsion system, which consists of engines and propellers. Sensors consist of inertial measurement unit (IMU) and inertial navigation system (INS) components, including rate gyros for roll, pitch and yaw motion, multi-axis accelerometers, digital compasses for directional information, pressure transducers for airspeed and altitude, ultrasonic range finders for measuring the distance to nearby objects, and electro-optical (EO) and infrared (IR) cameras. The guidance, navigation and control (GNC) system, the estimation/filtering system, and the health management (HM) scheme run on the flight computers. The transmitters and receivers (Tx/Rx) are

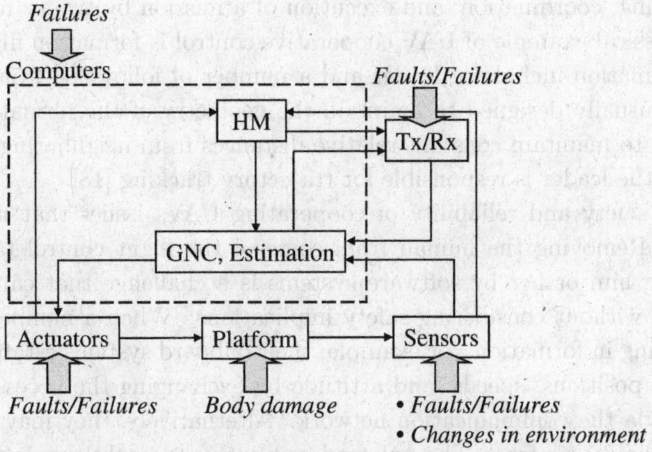

Fig. 1.1 Simplified schematics of UAS control system.

connected to the flight computers. Data obtained from Rx and the sensors are processed by the flight computers, which drive the actuators and the Tx to steer the vehicle accordingly and to transmit relevant information to the rest of the team and the operating crew.

Unmanned aircraft systems can be found in several sizes and exhibit various degrees of autonomy. A radio-controlled aircraft has the simplest level of autonomy, while an autonomous swarm exhibits the highest level of autonomy [9]. Increasing the autonomy of unmanned platforms could reduce the number of operators per vehicle, thus simplifying the task of the operating crew controlling vehicles involved in complex missions and potentially reducing costs. This book focuses on small-scale and miniature or micro UAVs, referred to respectively as SUAVs and MAVs [3, 9–16]. The acronyms UAV and UAS as used here refer to small-scale, miniature or micro unmanned aerial platforms with onboard and offboard systems dedicated to flight control and to accomplishing a mission. Briefly, these UAVs can fly close to the ground in confined areas and vary in size from a few meters to a few centimeters.

1.2 Cooperative Control

"Cooperative control" refers to a group of dynamic entities exchanging information to accomplish a common objective [17]. Cooperative control entails planning, coordination, and execution of a mission by two or more UAVs. A classical example of UAV cooperative control is formation flight. A typical formation includes a leader and a number of followers. Control schemes are usually designed to maintain the geometry of the formation. Followers try to maintain constant relative distances from neighboring vehicles, while the leader is responsible for trajectory tracking [18].

Why are safety and reliability of cooperating UAVs issues that need addressing? Removing the human from some of the flight control tasks and replacing him or her by software systems is a challenge that cannot be addressed without considering safety implications. When a number of UAVs are flying in formation, for example, their onboard systems establish their relative positions, speeds, and attitude by exchanging the necessary information via the communication network. Alternatively, they may use onboard proximity sensors. The onboard computers, namely the control systems, then use this information to produce a cohesive flight. Suppose one of the actuators of a UAV in the formation develops a fault. If the control system of the faulty UAV is not equipped with some form of robustness to fault or fault tolerance, or if the fault-tolerant control system is not capable of providing sufficient recovery to the fault, the vehicle may lose stability and exhibit an unpredictable pattern. As the control systems of the UAVs flying in formation aim at maintaining certain relative distances, velocities, and attitude at prescribed values under nominal conditions, the stability and cohesiveness of the formation may be lost. If their control systems are designed for nominal operating conditions, when the leader vehicle is at fault the follower vehicles will simply follow in its tracks without compensating for its erratic trajectory. Unless some sort of fault tolerance is embedded in the individual UAV GNC system and in the multi-UAV cooperative control system, the mission may be lost. Faulty aerial vehicles, and those naively following them, become inefficient in terms of energy consumption, fail to fulfil mission objectives, and represent a danger to humans.

Figures 1.2 and 1.3 illustrate two examples of cooperative control. In Fig. 1.2, a group of UAVs fly in a string-like formation. Cohesive group flight is ensured as follows: the control system of follower vehicle 1 (F1) acts to maintain a relative separation from the leader (L), and the control system of follower 2 (F2) does the same with respect to F1. Information flows from L to F1, and from F1 to F2. Information consists of d_L and v_L, representing relative distance and velocity between the leader and follower F1. This information is obtained through a communication network or from onboard proximity sensors and processing. The feedback control system onboard each vehicle uses inter-vehicle information exchange to compare relative distances and velocities and takes corrective action to maintain them at prescribed values. The bottom of Fig. 1.2 is a block diagram of the feedback loops showing the interdependence among the feedback control systems. The formation flight problem is discussed in more details in Chapter 3.

Figure 1.3 presents an example of cooperative control and decision making. Three formations of small UAVs plan their paths to coordinate rendezvous on three targets in a constrained and hostile environment. The grid represents the streets of a city. Starting from the base, the UAVs fly at low altitude and are thus constrained by nearby structures. The formations are shown at three time instants, from time t_1 to time t_3. A square represents a threat in the sense that the safety of a UAV is at risk along a path leg

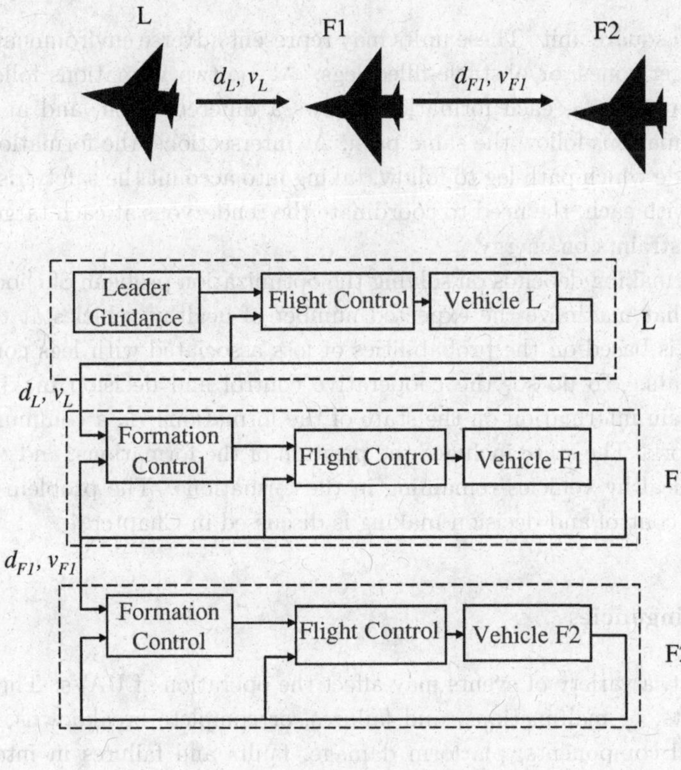

Fig. 1.2 Formation flight.

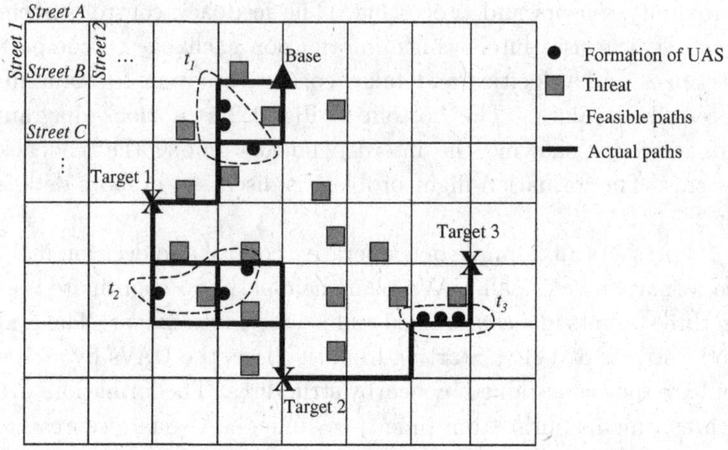

Fig. 1.3 Cooperative path planning and rendezvous.

comprising a square unit. These units may represent adverse environmental effects, danger zones, or obstacle-filled legs. At t_1, two formations follow the same path, at t_2, each formation follows a different path, and at t_3 all three formations follow the same path. At intersections, the formations have to decide which path leg to follow, taking into account the safety risks associated with each, the need to coordinate the rendezvous at each target, and the constraints on energy.

Decision making depends on solving the optimization problem of choosing routes that maximize the expected number of healthy vehicles at the targets and is based on the probabilities of loss associated with legs comprising threats. To do so, the cooperative control and decision making systems obtain information on the state of the formations via a communication network. The state includes the position of the formations, and the number of healthy vehicles remaining in the formations. The problem of cooperative control and decision making is discussed in Chapter 4.

1.3 Contingencies

During flight, a variety of events may affect the operation of UAVs. These include faults, or malfunctions, and failures, or complete breakdowns, in flight-critical components, platform damage, faults and failures in intervehicle information flow, anomalous behaviors or environmental occurrences such as bursts of wind, extreme weather, or icing on the airframe. Certain events are more likely to occur than others, depending on the context, and several different contingencies may be encountered concurrently. Furthermore, one contingency may lead to another. For example, cold weather may lead to control surfaces freezing and not responding as expected. Safe and reliable multi-UAV operations require systems that can handle such contingencies, as there are many off-nominal conditions that humans cannot handle in a timely and effective manner. It is not the purpose of this book to study ways of recovering from faults and failures affecting individual vehicle flight-critical components, software and systems. It is rather to present a number of UAV team cooperative monitoring and adaptation techniques and algorithms for a set of degraded conditions, building upon basic principles.

1.3.1 *Faults and failures of UAV components*

Faults and failures in UAV flight-critical components include those affecting sensors, actuators, flight computers, engine, and control surfaces. Faults in components in the control loop, as in Fig. 1.1, may compromise UAV flight. Such faults are known as component-level (CL) faults. This book considers actuator, control effector and sensor faults.

Common faults include the actuator or control effector getting stuck in a certain position and not responding to commands, the actuator having lost its authority, the actuator or control effector moving to its upper or lower limit, and the gain of the actuator becoming a fraction of its nominal value [19, 20]. For example, if the control surfaces of a fixed-wing UAV (such as the aileron, rudder or elevator) get stuck, they may stop responding to actuator commands or may only partially respond to commands. The consequence of a control surface fault is reduced performance and possibly instability, depending on the effectiveness of the health management system. A fault is distinct from a failure in that a fault is a malfunction, whereas a failure suggests complete breakdown of a system component or function [21].

Sensors in the UAV feedback loop in Fig. 1.1 are subject to both hardover failures, which are catastrophic but relatively easy to detect, and soft failures, which are difficult to detect but nonetheless critical [22]. Hardover failures are typically detected and identified by a sensor with built-in testing. Soft failures include a small bias in measurements, slow-drifting of measurements, a combination of the two, loss of accuracy, and freezing of the sensor to certain values [23]. For UAVs equipped with GPS receivers, examples of faults include jamming of GPS data (intentionally or not) and the multi-path effect of reflections causing delays. These in turn result in inaccurate positions, and can have important consequences in dense urban terrain [24]. Sensors used for vision feedback may also suffer from failures [25]. A fault in a sensor alters the measurements required by controllers, and depending on the severity of the fault, may degrade the closed-loop performance.

1.3.2 *Vehicle damage*

An environmental hazard may cause damage to a UAV in areas of high density. The impact of the damage on UAV performance depends on the severity of the damage and on the effectiveness of the health management system. The platform itself may be damaged as well as the flight-critical components and systems. The partial destruction of an actuator, sensor or flight computer during flight may be interpreted as a CL fault.

In the case of a fixed-wing UAV, for example, control surface damage can change the dynamics, translating in a modified control input-to-state matrix and as an additional nonlinear term representing asymmetrical postdamage dynamics [19, 26]. If the dynamics of the vehicle are radically changed, the control system may need to employ online system identification and adaptation techniques and re-allocate the control effort to the remaining control surfaces to preserve a certain level of performance. Reference [27] proposes to model the body damage of an airship-type UAV as a change in the buoyancy force.

1.3.3 *Information flow faults*

Inter-vehicle communications are needed in any collaborative effort. Mobile ad hoc networks enable wireless transmission of data in dynamic environments over radio waves. The topology of these computer networks may vary with time, with nodes joining and leaving the network depending on their distance from one another. IEEE 802.11 standards are widely used with off-the-shelf computer network technology. Each UAV can be viewed as a node equipped with wireless Tx/Rx capable of transmitting and receiving data packets to and from its neighbors. The wireless medium is, however, unreliable. Wireless communications are subject to environmental intrusions that interfere with the signals and block their paths, introducing echoes, noise, and jamming [28]. The limited available onboard power, the mobility of the vehicle, and the presence of nearby users also constrain the effectiveness of inter-vehicle communications [24, 29].

An information flow fault is a temporary or permanent loss of information between two or more UAVs, or between UAVs and the operating crew. This type of fault may affect the wireless network medium or the alternative means of communication, such as those relying on sensor data. The information flow fault may arise from a communication breakdown due to obstacles, jamming, a node loss, the crash of a UAV, or because of Tx/Rx or flight computer failures, for example. If the loss of information is permanent and affects all forms of inter-vehicle communications concurrently, cooperation cannot be reestablished among the vehicles. Figure 1.4 illustrates the information flow among a team of unmanned vehicles for the purpose of carrying out a collaborative operation. Information can be transmitted by means of the wireless communication network, onboard sensors, and subsequent interpretation by onboard processing. Each vehicle requires information about its teammates, such as current position, velocity, health status, target zones visited, assigned targets, and so on. The information shared by the systems is specific to the particular mission, and the vehicles may or may not communicate with the ground crew. UAS flight computers comprise GNC and HM systems. The cooperative control system functions may be found onboard each UAS, offboard the vehicles, or onboard a subset of the UAS fleet.

Loss of some or all communication links during flight may occur for various reasons. A fault in the Tx/Rx devices themselves may result in an information flow fault. Sensors being used as an alternative means of communication to collect information on neighboring vehicles may also be at fault. Regardless of the cause of the information flow fault, elements of information are missing and, unless the health management system enables the UAVs to handle information loss, the integrity of the collaborative mission may be jeopardized.

1.3.4 *Team anomalies and collisions*

An anomaly may be loosely defined as a deviation from nominal behavior. Vehicles that stray away from their intended trajectory or from their expected motion exhibit anomalous behavior, increasing the risk of collision with the surroundings and with neighboring vehicles. Causes of erratic or anomalous vehicle motion might be actuator and sensor faults and failures,

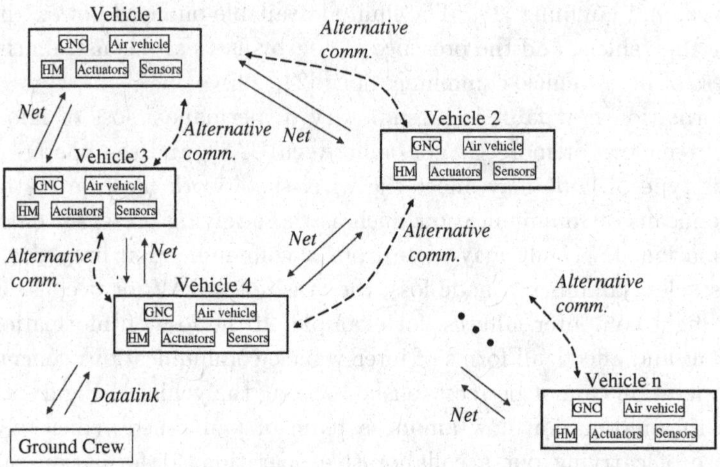

Fig. 1.4 Information flow in UAV team. Information is transferred by means of wireless communication network (Net) and alternative broadcast means (UAV motion + sensor + processing).

platform damage, or control of the UAV by hostile forces, for example. Preventing a trickle-down effect from the fault to the entire team requires a monitoring system capable of detecting anomalous behaviors relatively quickly, and an adaptation mechanism. Such a system would enable neighboring UAVs to avoid colliding with the anomalous vehicle while adapting to the contingency so as to continue their operation as effectively as possible. The prevention of collision is an important aspect of any safety system, and much work has been done on collision avoidance for unmanned vehicles. A classical approach to avoiding collisions among agents and between agents and obstacles is the use of potential fields and variants thereof [30, 31]. Another approach to collision-free navigation relies on model predictive control and is demonstrated experimentally in Ref. [32]. While we do not tackle this theme here, the health management system we propose does take into account the impact of collisions on UAVs. When UAVs collide, they are either lost or damaged, and the health management system accommodates these changes in the health status of the fleet.

1.3.5 *Environmental effects*

A degradation in the performance of UAV sensors and actuators may be the result of poor weather or other adverse environmental effects. The environment may include other vehicles (manned or unmanned), natural phenomena (winds, hurricanes, cold, heat, humidity, dryness, fires, sand, snow storms, and so on), man-made structures, animals, humans, or toxic substances, to cite a few. Forces of nature have a greater impact on small vehicles than on large vehicles. There are a number of reasons for this. In cases where precise positioning of the body is needed, for example, the aerodynamic scale and available control authority make it difficult for MAVs and SUAVs to counteract the effects of the wind. For larger UAVs, technologies similar to those used in military and commercial manned aircraft, designed to handle typical wind forces, can be readily employed. Furthermore, constraints on available onboard power, as well as allowed payload mass and volume, indirectly limit the complexity and sophistication of the control laws that can be embedded on small UAVs, and these factors necessarily restrict the choice and number of actuators and sensors. As a consequence, it is more difficult for small UAV systems to detect the signs and characteristics of certain weather conditions, and respond effectively.

1.3.6 *Book overview*

The main purpose of this book is to present techniques for designing health management systems for cooperating UAS, with objective of ensuring safety and reliability. Safety in this context is defined as reducing human exposure to risk, and reliability as enabling the successful completion of a cooperative mission despite unexpected events. Concepts of health management relating to individual UAS are not covered exhaustively, although we provide a brief overview of some of the main techniques employed for individual vehicle health management. Chapter 2 lists a panoply of existing methods for addressing component-level faults and failures in general dynamic systems, in particular those affecting actuators and sensors. The principal approaches to fault-tolerant control and fault/failure detection and diagnosis are described with reference to several studies.

The heart of this book is the presentation of cooperative health management techniques in Chapters 3 and 4. We believe that health management techniques are needed at the team level, for instance in cases where component-level faults and failures are not adequately handled by the individual vehicle's fault-tolerant control and fault/failure detection and diagnosis schemes. These two chapters discuss cooperative control synthesis, analysis, and implementation, as well as the design of health management for cooperating UAVs. Chapter 3 focuses in more detail on decentralized team monitoring and vehicle adaptation in cases where anomalous behaviors occur when UAVs are flying in formation. The chapter begins with a description of models, including dynamic models of both airship and quadrotor UAVs, flight control, and faults. The chapter goes on to present a formation flight control law, which relies on the concept of passivity [33] and on contraction theory [34]. Two approaches are proposed for health monitoring in formation-flying UAV teams: an observer-based abrupt fault detector and a signal-based non-abrupt fault detector. Abrupt faults are generally characterized by a sudden, stepwise change in the observed variables, while non-abrupt faults evolve relatively slowly. Simple rules are presented for adapting vehicles to the detection of faults. Chapter 3 ends with illustrative simulation and experimental results, demonstrating the performances obtained with the monitoring and adaptation system for single and mixed-type simultaneous faults; followed by a summary.

Chapter 4 proposes cooperative health management techniques integrated with decision making for a surveillance mission taking place in a dynamic and uncertain environment. A UAV team is tasked to reach a set of target locations. In moving from one location to another, the vehicles have to follow narrow paths, and may face health-threatening situations. UAS-threat encounters are modeled as Markov decision processes [35], relying on probabilistic models of UAV survival. Decision policies are sought such that the cooperative path planning of the UAS maximizes the overall, expected survival. The policies rely on dynamic programming and heuristic techniques [36]. The academic case of a perfectly known environment and the more practical case of a partially known environment are studied. Chapter 4 continues with the description of a method for handling communication failures. A health state estimator ensures effective cooperation among the UAVs despite intermittent wireless communication breakdowns. Chapter 4 provides extensive numerical simulations to demonstrate effectiveness and integration of the various functionalities of the cooperative health management and decision making system. Chapter 4 ends with approaches to the implementation of the algorithms and for the rapid prototyping of cooperative control systems.

Chapter 2

Health Management for the Individual Vehicle: A Review

An automatic system, referred to as the health management system, can be designed and embedded onboard an unmanned vehicle to reduce its vulnerability to faults and failures in the actuators, control surfaces and sensors. The term health is used in reference to individual UAV guidance, navigation, and control. A healthy UAV is defined as one that performs as expected under nominal conditions; that is, when unaffected by contingencies such as CL faults, platform damage, and environmental effects. The health management system monitors the health state of the systems and if there is detection of a fault, diagnoses the nature of the condition, provides corrective action within the feedback control loop, and ultimately commands the remaining control surfaces. According to the Merriam-Webster dictionary, diagnosis pertains to the investigation or the analysis of the cause or nature of a condition, situation, or problem. To minimize the effects of the fault on the vehicle, and hence minimize damage and loss of life, detection and diagnosis should ideally be carried out automatically, and early enough to allow for recovery, or reconfiguration, of the control system. In short, the health management system must be able to detect the anomaly and select the appropriate method of remedy or recovery. For relatively large vehicles, this is achieved through hardware redundancy in combination with software. This approach assumes that redundant actuators and sensors are present onboard the vehicle, and hardware units may be duplicated and triplicated. Such physical redundancy entails larger mass, increased maintenance, additional cost, and a more complex system.

Using more sophisticated software to perform the health management functions without resorting to duplication of channels and hardware is an alternative approach to relying on physical redundancy. Techniques using this approach make use of mathematical models of the dynamics and flight-critical components of the vehicle, relying on the relationships between input and output of sensors and actuators and leveraging any available redundancy of systems. Software-based health management offers flexibility, growth capability, and wide applicability. Individual vehicle health management integrated with the control system can be designed to detect faults, identify their origin and enable recovery, thus providing a certain level of robustness to faults and failures. This system is commonly known as a fault-tolerant control (FTC) system. Research into fault-tolerant control has been expanding for a number of years, and the techniques developed have been used in a wide range of applications. Before presenting current approaches to UAV health management, it is therefore worth reviewing the main approaches to FTC for aircraft.

2.1 Passive and Active Fault-Tolerant Control Systems

FTC systems can be described as control systems that maintain closed-loop stability and acceptable transient and steady-state performance regardless of CL faults and failures [21, 37]. For the UAV, FTC systems aim at maintaining stable flight and a certain level of performance in spite of onboard actuator and sensor faults as well as body damage. To achieve this, however, the aircraft must contain a sufficient number of redundant actuators and sensors. For example, in the case of a small delta-canard fighter aircraft configuration, as simulated by ADMIRE [38], there are seven control surfaces (not counting flaps, air brakes, and thrust vectoring): right and left canards, right outer and inner elevons, left inner and outer elevons, and rudder. This configuration provides a high degree of redundancy, which is leveraged by the FTC system.

There are two types of FTC systems: passive and active. A passive FTC system can accommodate certain limited types of CL fault without needing online information on the health state of the control system. It can maintain a certain level of performance because it is designed for a set of known, anticipated faults, and is implemented using fixed parameters. This approach can be defined as a robust control design. There are a variety of techniques for designing robust controllers [39]. Some design techniques result in adequate closed-loop performance despite faults [40]. This kind of robust controller is sometimes referred to as a reliable controller [41, 42]. Figure 2.1(a) uses typical block diagrams to illustrate the passive FTC system. Robust control design should address exogenous disturbances, system uncertainties, faults, failures, and sensor noise.

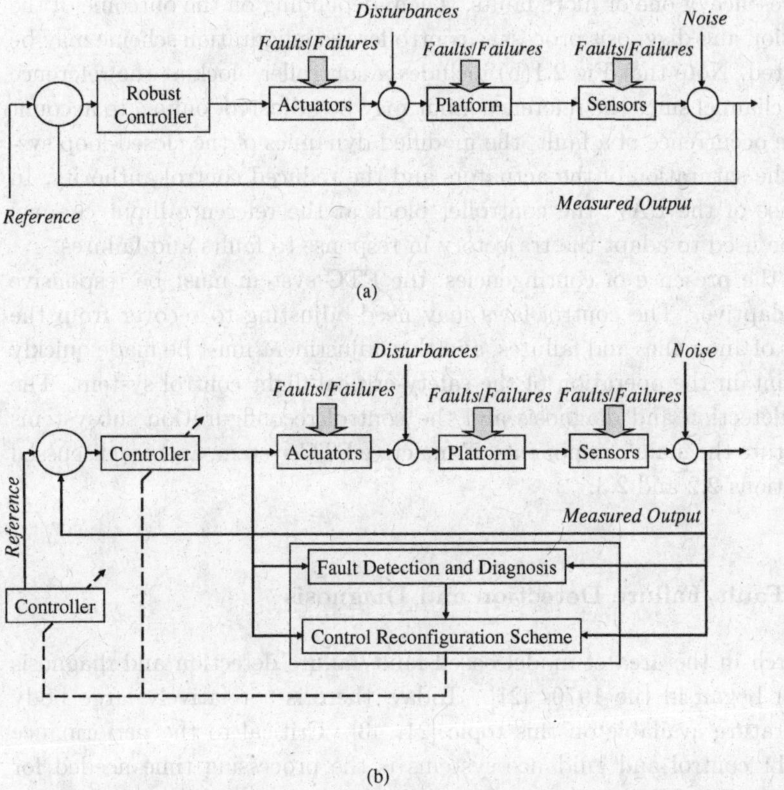

Fig. 2.1 Block diagrams of (a) passive FTC, and (b) active FTC.

An active FTC system, on the other hand, accommodates CL faults by reconfiguring the control law online, and in real time, based on the most up-to-date information on the control system. Active FTC requires fault detection and diagnosis, as well as controller reconfiguration (recovery). Unlike passive FTC, active FTC can compensate for unanticipated faults. Fault detection, diagnosis and reconfiguration represent the main components of an active FTC system, as shown in Figure 2.1(b). The acronym FDIR is often used to refer to the area of research devoted to fault detection, isolation and recovery. Performing fault isolation is one of the basic diagnosis processes. The FTC system uses the control input, the measured output, and available mathematical models to detect and diagnose the presence of one or more faults. Then, depending on the outcome of the detection and diagnosis process, a controller reconfiguration scheme may be activated. Note that Fig 2.1(b) includes a controller block at the reference input channel since the reference input may be adjusted, online, to account for the occurrence of a fault, the modified dynamics of the closed-loop system, the saturation of the actuators and the reduced control authority. In the case of the UAV, the controller block at the reference input channel may be used to adapt the trajectory in response to faults and failures.

In the presence of contingencies, the FTC system must be responsive and adaptive. The control laws may need adjusting to recover from the effects of anomalies and failures, and this adjustment must be made quickly to maintain the operation of the safety-critical flight control system. The fault detection and diagnosis and the control reconfiguration subsystems constitute the main components of an active FTC system, and are discussed in Sections 2.2 and 2.3.

2.2 Fault/Failure Detection and Diagnosis

Research in the area of model-based fault/failure detection and diagnosis (FDD) began in the 1970s [21]. Today, there is a relatively large body of literature available on this topic [21, 43]. Critical to the performance of flight control and guidance systems is the processing time needed for the controller, FDD, and reconfiguration subsystems in the feedback loop. Among the techniques available for FDD, those requiring relatively little processing time are preferable, as this will ensure that computations can take place within hard real-time deadlines. In addition to the timing issue, the closed-loop FTC system should aim at providing robust closed-loop performance to system uncertainties while preserving FDD subsystem sensitivity to faults. Robust online FDD systems first detect and isolate faults/failures of the actuators and control surfaces, and then reconfigure the guidance and control laws accordingly [44]. Several schemes limited to the isolation of faults are referred to as fault detection and isolation (FDI) methods. Common model-based FDD schemes usually consist of two phases: (1) residuals are generated by a filter or an observer, and (2) residuals are analyzed to decide whether a fault has actually occurred and, if so, to isolate the actuator affected by the fault. The residuals should be sensitive to the occurrence of faults, and should contain enough information to enable the FDD system to both discriminate among the various types of faults and between healthy and faulty actuators, regardless of disturbances, noise, and parameter variations.

Over the past decades, a number of model-based FDD methods have been developed [45, 46]. Techniques for actuator and sensor FDD include observers [47], multiple models with consideration of closed-loop performance degradation [44, 48], multiple-model adaptive estimation [49], parity space methods systematically accounting for uncertainties [45, 50, 51], and parameter and state estimation [45, 52]. Some robust fault detection techniques are based on eigenstructure assignment [53], or on eigenstructure assignment combined with linear matrix inequalities to reduce the impact of disturbance and to reduce computation time [54], or on sliding modes [55]. Unknown input observers can also be employed for robust FDD. Such techniques aim at decoupling the state estimation errors from the disturbances affecting the system, the so-called unknown inputs [21].

Most model-based FDD techniques are devised for linear systems or linear models of nonlinear systems. FDD methods designed for nonlinear systems have the advantage of avoiding the linearization phase, or having to approximate the dynamic system. Nonlinear FDD techniques include unknown input observers [56], the geometric approach, which uses coordinate transformation to provide necessary and sufficient conditions for solving the problem of detecting and isolating faults in general nonlinear systems [57], the robust observer [19], and adaptive and online approximation approaches based on learning techniques [58, 59].

In the decision phase of common model-based FDD techniques, a comparison is made between the signals of interest, namely the residuals, and the threshold. The threshold corresponds to the divide between acceptable (healthy) and unacceptable (faulty) behaviors. Choosing the value of the threshold is a difficult and critical task. Thresholds can be set by carrying out extensive simulations of a variety of flight scenarios and then performing online parameter scheduling, depending on the flight conditions. Thresholds can be adapted online to reduce the computational demand [60].

2.3 Control Reconfiguration

In active FTC systems, control reconfiguration modifies controllers by changing their structure and parameters using the information provided by the fault detection and diagnosis subsystem. However, some researchers make a distinction between restructurable control systems, which may have their structure altered online, and reconfigurable control systems, which may have their parameters changed online [61]. Controllers can be designed either (1) offline, to accommodate all of the expected faulty conditions, or (2) offline for the nominal conditions, and then re-designed online when faults are detected.

In approach (1), a set of controllers is designed prior to flight. Controllers are designed offline for the various nominal operating modes, e.g. takeoff, landing, and specific maneuvers, as well as for the anticipated faulty conditions. During flight, a switching mechanism triggers the control law corresponding to the current flight mode, which can be either a nominal condition or a faulty condition.

Approach (2), or online controller re-design, can be done by relying on feedback control concepts such as linear quadratic regulation [62], dynamic inversion [63, 64], model predictive control [65, 66], intelligent control [67–69], optimal control [70], eigenstructure assignment [71, 72], model following [73], and adaptive control [74–76], to name a few. The reconfigured control law leverages any redundancy in control effectors. For example, in the X-33 aircraft, eight independently actuated control surfaces are available: right and left rudders, body flaps, and inboard and outboard elevons [77]. This configuration provides redundant generation of pitch, roll, and yaw moments. Once a fault has been detected and possibly identified, the problem becomes how to use the remaining actuators and control surfaces to generate the required control inputs (moments and forces) [77, 78]. Depending on the reconfiguration method used, the FDD module may not be needed [79].

One of the challenges with online reconfiguration is the fact the control system may be faced with several transitions during flight. It is well known that such transitions can give rise to transients which may destabilize the aircraft closed-loop system. An abrupt transition, or switch, between controllers can excite high-frequency dynamics, causing undesired responses and stress on the actuators [80]. Plant states may change abruptly due to a fault (e.g. body damage) and the system may resort to an automaton to reconfigure the controller [81]. Interpreting the design of an active FTC system from a hybrid system framework is one possible design approach. A hybrid system is one that has both continuous and discrete states. Approaches ensuring relatively smooth transitions include those that rely on transition management [80], blending [39, 82], and state initialization [80, 83], to name a few.

2.4 FTC and FDD Techniques for MAV and SUAV

Designing guidance and control systems for small and miniature unmanned aerial vehicles is challenging due to inherent constraints in size, weight, computing power, and energy. Furthermore, in real operations the vehicles are subject to high vibrations, are exposed to the elements, especially the wind [84], are prone to temporary degradation, such as loss of GPS signals and radio communications, and are vulnerable to magnetic interference [24]. Payload constraints impose a fundamental limit to the quality of the sensors and actuators that can be placed onboard. Designing control, guidance and navigation systems ensuring functionality of an ailing vehicle requires efficient use of software to ensure adequate performance [85, 86].

There are some results available in the literature in the area of health management for small and miniature UAVs. First, on the experimental front, some research has been done on centralized health management of multi-vehicle systems, including the use of a health feedback loop in a receding horizon task assignment algorithm [85]; and on long-duration, multi-UAV autonomous mission-planning experiments incorporating vehicle health state [87]. Second, on the subject of system architecture, Ref. [88] presents a four-layer autonomous intelligent control system architecture for UAVs spanning the lower FDI level to the upper decision making level. Third, further research addresses control surface bias and the practical issues inherent to low-cost sensors and actuators [89]. A nonlinear observer is integrated to the autopilot of a COTS fixed-wing UAV to compensate for small aileron and elevator bias. Despite low-end sensor suites, as found in commercial radio-controlled, fixed-wing aircraft, where noise is significant, flight experiments showed that real-time reconfigurable control can be implemented on such vehicles with some degree of effectiveness [90]. Fourth, for the coordination of multiple vehicles, path planning and task assignment for a group of MAVs subject to wind, and potentially to failure of the assignment algorithm, were experimentally verified in Ref. [4]. Finally, as part of the DARPA Software Enabled Control Program, researchers studied fault-tolerant control of an unmanned rotorcraft and showed the effectiveness of the FTC system under a stuck collective pitch fault, as described in Ref. [91].

Chapter 3

Health Monitoring and Adaptation for UAS Formations

Consider a number of unmanned aerial vehicles flying together as a group. For instance, Fig. 3.1 illustrates a V-shape formation of six vehicles with a single leader at the front (L), and five follower vehicles (F1 to F5). The dashed lines indicate the information flow needed for the control of the formation; for example, they might indicate transmission of positions and speeds from one vehicle control system to another. The goal of a safe and reliable formation flight control scheme is to ensure cohesive flight and issue commands for specific formation geometries and vehicle attitudes, regardless of wind turbulence, aerodynamic effects between nearby vehicles, flight path constraints, obstacles, faults and failures of one or more actuators and sensors, and information flow faults.

Typically, formation geometry and vehicle attitude are set according to mission objectives and vehicle capabilities. For example, to ensure complete coverage of a certain area, UAVs flying at a constant altitude deploy their onboard sensors in particular orientations and maintain a specific geometry and body attitude over the zone being monitored. In case of constrained flight paths, such as in low-altitude flight through urban terrain, changes in the geometry of the formation are needed, as shown in Fig. 3.2 for a V-shape formation of six vehicles. The formation control system uses the communication network to pass messages on the commanded geometry, sending the required relative distances and velocities among the neighboring vehicles, as well as the prescribed orientations. Changing from one shape to another gives rise to transients in the motion of the vehicles. The formation must be such that the vehicles do not collide during such geometry changes.

One approach to achieving reliable leader-to-follower formations in the presence of actuator faults and failures is to design individual vehicle FTC and FDD systems. Although CL FTC and FDD techniques have been

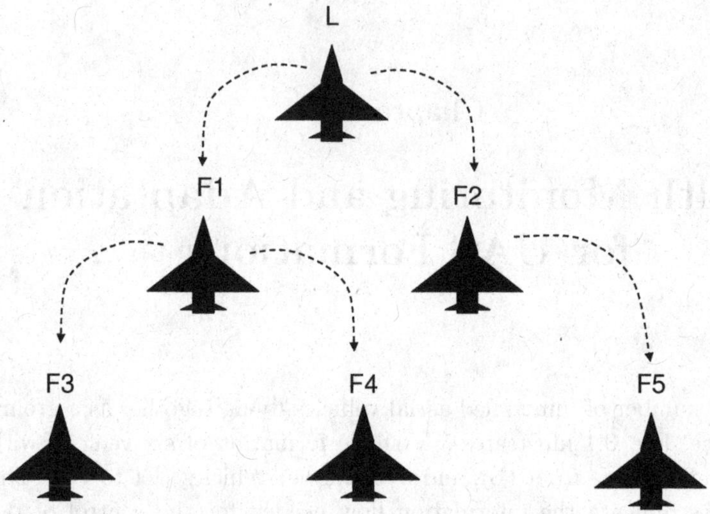

Fig. 3.1 A leader-to-follower formation of six vehicles.

extensively studied for individual aircraft, much remains to be done for formation-flying MAVs and SUAVs. For distributed formation control schemes, the trajectory of any follower vehicle in the formation depends on the motion of its predecessors. If a CL FTC system onboard an unmanned vehicle cannot preserve flight performance at an acceptable level when faced with certain faults or failures, the vehicle may undergo abnormal motion, which may lead to formation instability and collisions. Propagation of the fault to the rest of the formation can clearly be prevented if the fault detection can be communicated to the team members' control systems early enough, and if a team-level health management system is able to mitigate the impact of the fault on the team. However, if there is an information flow fault taking place in addition to the CL fault, health information cannot be passed to the rest of the formation. This potentially harmful situation motivates the design of a decentralized fault detection and command adaptation (DFD-CA) capability, where each vehicle can detect anomalous behaviors in neighboring vehicles and adapt its own commands accordingly, with the objective of preserving team goals. In other words, using available redundant information channels, such as onboard sensors, as well as information obtained from other manned and unmanned platforms and from

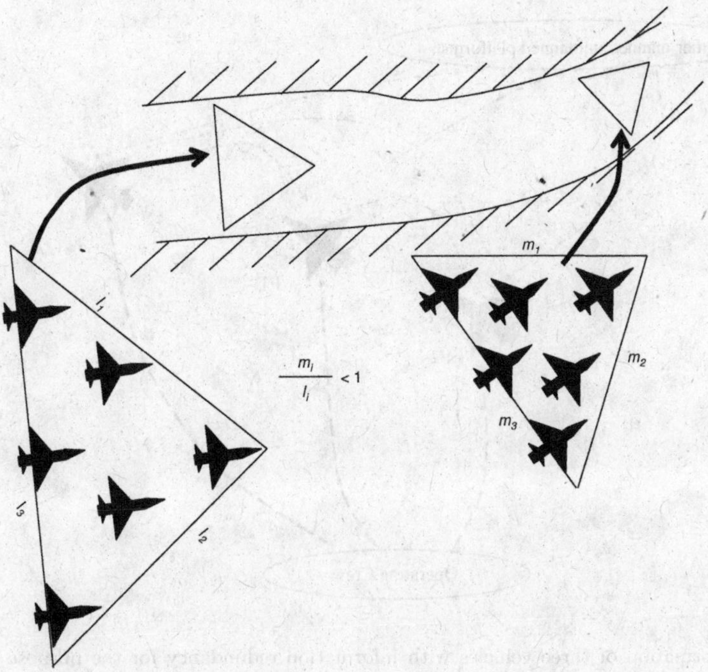

Fig. 3.2 Formation flight in a constrained environment. The formation geometry goes from shape 1 (l_i segments) to shape 2 (m_i segments).

the operating crew, a follower vehicle can monitor neighbors within range, as shown in Fig. 3.3. Dashed lines represent information flow, and the different types of dashed line indicate different communication rates. Once the measurements and other information is collected, the DFD-CA scheme determines whether the vehicle should continue following its predecessor or follow another team member within range. While individual vehicle FTC and FDD are concerned with self-monitoring, team health management seeks to increase awareness of the environment and of the neighbors.

This chapter focuses on two techniques for DFD-CA of aircraft flying in formation. One relies on the design of an observer, whereas the other compares key signals. Before these two techniques are discussed, however, Sec. 3.1 introduces mathematical models of UAV dynamics, individual vehicle flight control schemes, actuator faults, and information flow faults. The formation geometry is ensured by means of a formation flight control algorithm derived for nominal, non-degraded conditions of operation, and

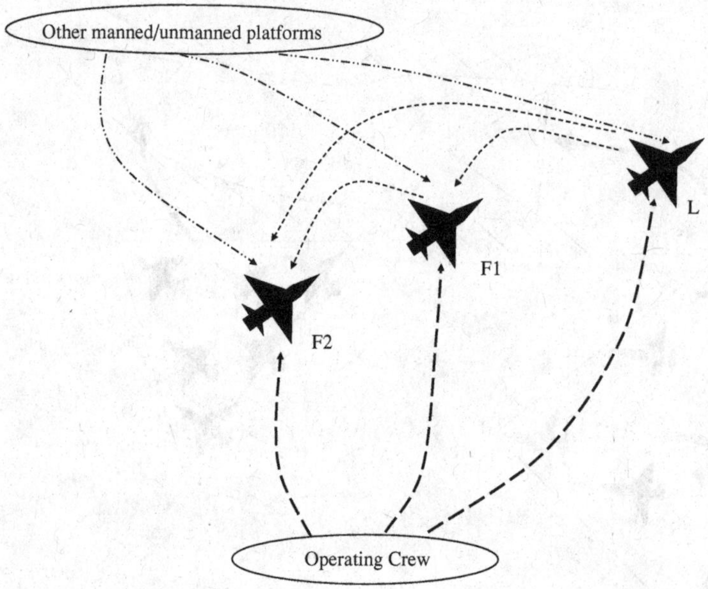

Fig. 3.3 Formation of three vehicles with information redundancy for the purpose of health management.

presented in Sec. 3.2. Algorithms that complement the formation control scheme by handling component-level faults from a team perspective are presented in Secs. 3.3 and 3.4. A CL fault may significantly perturb the performance of a vehicle, for example, if the local FTC-FDD system provides limited recovery or is subject to a malfunction, or if the vehicle simply does not possess such a system. Section 3.3 details the observer-based, abrupt fault detector, whereas the non-abrupt fault detector is described in Sec. 3.4. Section 3.5 describes the adaptation mechanism. Together, Secs. 3.2 to 3.5 provide a set of complementary techniques for DFD-CA of formation flying UAVs. Indeed, the health monitoring system relies on information redundancy to feed an observer and a signal comparator, and to address faults characterized by fast (abrupt) and slow (non-abrupt) dynamics. Figure 3.4 shows the schematics of the formation control and health management functions, whose computing may take place onboard or offboard the UAV depending on the available processing power. The last section of the chapter describes the results and analysis of numerical simulations and experiments pertaining to the DFD-CA algorithms. The

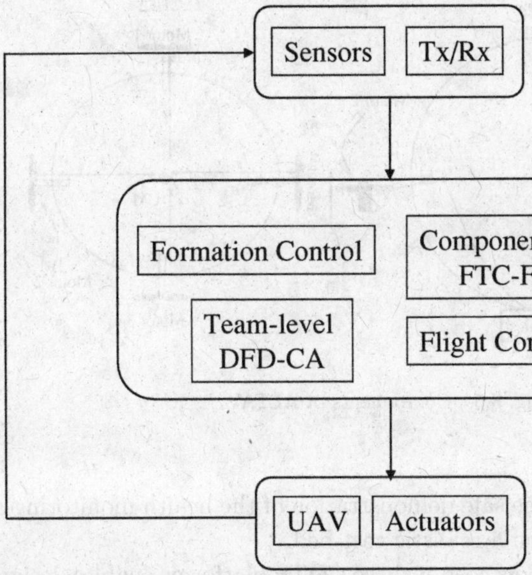

Fig. 3.4 Schematics of decentralized fault detection and command adaptation functions integrated with control functions.

section includes a comprehensive discussion of the performances obtained with formation monitoring and adaptation in case of single and mixed-type concurrent component-level and information flow faults.

3.1 Models of Vehicle Dynamics, Flight Control, and Faults

The mathematical models of two UAVs are presented: an almost-lighter-than-air vehicle (ALTAV), and a quadrotor. These vehicles are selected for three reasons.

First, such small-scale vehicles have limited payload capacity and control actuation, making the design of their guidance, navigation, control and health management systems particularly challenging. Computing resources for running the various control functions are limited. Also, hardware redundancy is practically nonexistent.

Second, these vehicles facilitate emulation of situations in which vehicles are at fault. Their relatively slow dynamics and their smooth degradation of performance when faced with actuator faults, pertinent to the ALTAV

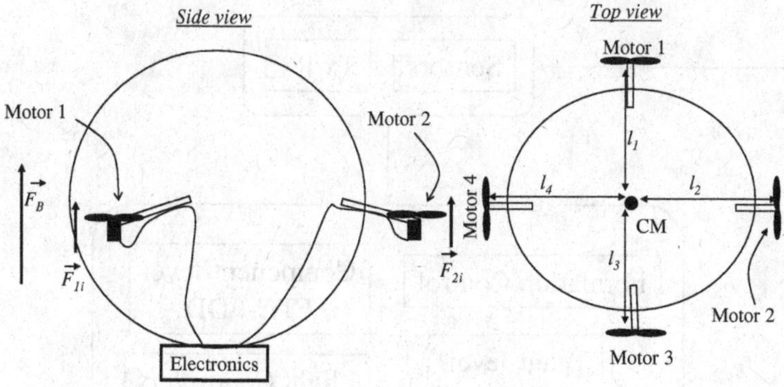

Fig. 3.5 Components of ALTAV.

in particular, allow for the safe demonstration of the health monitoring and adaptation capabilities on a realistic test-bed.

Third, the relatively low cost of these COTS platforms enables designers to test their control systems designs with limited personnel and resources, speeding up the prototyping time.

3.1.1 *ALTAV dynamics and control*

The ALTAV is shown in Fig. 3.5. The vehicle is approximately 1.5 m in diameter. The lift is generated by four DC motors and by the buoyancy of the helium-filled balloon. The dynamics of the ith ALTAV in a formation is governed by the following model [14, 15]

$$\begin{aligned}
M\tfrac{d^2x_i}{dt^2} &= \textstyle\sum_j F_{ji}\sin(\gamma_i) - C_x\tfrac{dx_i}{dt}, \\
M\tfrac{d^2y_i}{dt^2} &= \textstyle\sum_j F_{ji}\sin(\phi_i) - C_y\tfrac{dy_i}{dt}, \\
M\tfrac{d^2z_i}{dt^2} &= -\textstyle\sum_j F_{ji}\cos(\gamma_i)\cos(\phi_i) - F_B + F_g - C_z\tfrac{dz_i}{dt}, \\
J_\theta\tfrac{d^2\theta_i}{dt^2} &= (F_{1i}l_1 - F_{2i}l_2 + F_{3i}l_3 - F_{4i}l_4)\sin(\rho_i) - C_\theta\tfrac{d\theta_i}{dt}, \\
J_\gamma\tfrac{d^2\gamma_i}{dt^2} &= F_{1i}l_1 - F_{3i}l_3 - F_Bl_B\sin(\gamma_i) - C_\gamma\tfrac{d\gamma_i}{dt}, \\
J_\phi\tfrac{d^2\phi_i}{dt^2} &= -F_{2i}l_2 + F_{4i}l_4 - F_Bl_B\sin(\phi_i) - C_\phi\tfrac{d\phi_i}{dt}.
\end{aligned} \tag{3.1}$$

In Eqs. (3.1), x_i, y_i, and z_i represent vehicle translations in inertial frame; θ_i, ϕ_i, and γ_i are the vehicle rotation angles; M is the mass of the

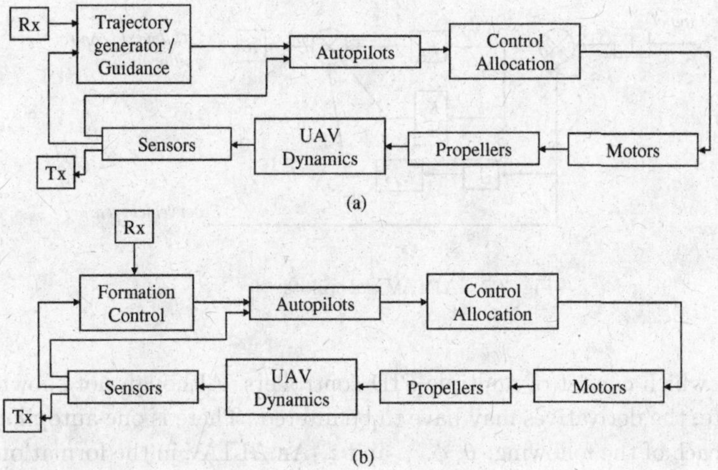

Fig. 3.6 Guidance and control loops for (a) leader vehicle, and (b) follower vehicles.

vehicle; J_ϕ, J_γ, and J_θ are the moments of inertia about the x, y, and z axes attached to body; F_g is the force due to gravity; F_B is the buoyant force resulting from the volume of helium; F_{ji} is the magnitude of the force of the jth motor, $j \in \{1, 2, 3, 4\}$; l_j is the distance from a motor to the center of mass (CM) of the vehicle, perpendicular to the direction of the force generated by the motor, $j \in \{1, 2, 3, 4\}$; C_i is the drag coefficient in the directions $i \in \{x, y, z, \theta, \phi, \gamma\}$, which serves as a damping term for the motion in that direction; and ρ_i is the angular offset from the vertical axis of the motor thrust vectors (around 5 degrees) for yaw control [14, 15]. The mathematical model (3.1) ignores the coupling effects, although it provides sufficient detail for the material presented here.

The flight control and guidance laws for the leader and follower ALTAVs are shown schematically in Fig. 3.6. In part (a), the control loop of the leader vehicle is shown. The trajectory of the leader vehicle may be obtained by means of offboard or onboard computing. Receiver devices enable communications with the operating crew. Thus, changes to the trajectory or to parameters of the mission may be communicated to the leader vehicle. In part (b) of Fig. 3.6, the control loop of the follower vehicles is illustrated. The formation control block feeds the autopilot block. The calculated command forces are sent to the control allocation function that drives the inputs to the four motors. Figure 3.7 shows the generic ALTAV

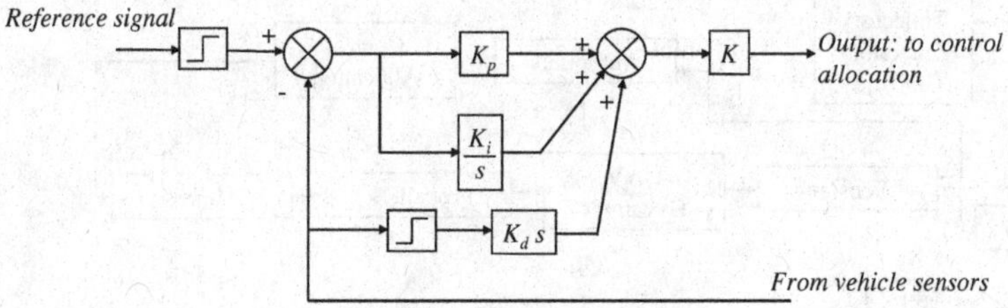

Fig. 3.7 ALTAV autopilots.

autopilots, which consist of multiple PID controllers. Although not shown in the figure, the derivatives may have to be filtered. There is one autopilot to control each of the following: θ, ϕ, γ, and z. An ALTAV in the formation is identified with subscript i. For leader-to-follower formations, the leader has subscript 0, whereas the followers are identified with a positive integer subscript. The autopilot and control allocation subsystems provide commands to the four motors to maintain the specified altitude, yaw, pitch and roll. The forces obtained can be expressed as follows for the ith ALTAV:

$$\begin{aligned}
F_{1i} &= Sat_{0,S_F}(f_i + g_i + \lambda_i),\\
F_{2i} &= Sat_{0,S_F}(-h_i^c - g_i + \lambda_i),\\
F_{3i} &= Sat_{0,S_F}(-f_i + g_i + \lambda_i),\\
F_{4i} &= Sat_{0,S_F}(h_i^c - g_i + \lambda_i),\\
f_i &= p_{\varphi\gamma}(Sat_{-S_{\varphi\gamma},S_{\varphi\gamma}}(u_{xi}) - \gamma_i)\\
&+ i_{\varphi\gamma}\int_{t_i}^{t}(Sat_{-S_{\varphi\gamma},S_{\varphi\gamma}}(u_{xi}) - \gamma_i)dv - d_{\varphi\gamma}\dot{\gamma}_i,\\
h_i^c &= p_{\varphi\gamma}(Sat_{-S_{\varphi\gamma},S_{\varphi\gamma}}(u_{yi}) - \varphi_i)\\
&+ i_{\varphi\gamma}\int_{t_i}^{t}(Sat_{-S_{\varphi\gamma},S_{\varphi\gamma}}(u_{yi}) - \varphi_i)dv - d_{\varphi\gamma}\dot{\varphi}_i,\\
g_i &= -p_\theta\theta_i - i_\theta\int_{t_i}^{t}\theta_i dv - d_\theta\dot{\theta}_i,\\
\lambda_i &= p_z(z_i - z_t) + i_z\int_{t_i}^{t}(z_i - z_t)dv - d_z(\dot{z}_i - \dot{z}_t)
\end{aligned} \tag{3.2}$$

where $p_{\varphi\gamma}$ (proportional), $i_{\varphi\gamma}$ (integral), and $d_{\varphi\gamma}$ (derivative) are the gains for both PID autopilot laws f_i and h_i^c; p_θ, i_θ, and d_θ are those of g_i; and p_z, i_z, and d_z are the gains of the PID altitude autopilot law. Signals u_{xi} and u_{yi} are generated by the outer-loop control law to stabilize the formation on the $x - y$ plane along a trajectory. Follower control laws u_{xi} and u_{yi} are given in detail in Sec. 3.2. In Eqs. (3.2), the saturation function of an actuator is given as

$$Sat_{a,b}(x) = \begin{cases} x, & a \le x \le b \\ b, & x > b \\ a, & x < a. \end{cases} \tag{3.3}$$

In Cartesian coordinates, outputs of the leader autopilots can be expressed as follows. The flight controllers generate pitch and roll commands (u_x, u_y) to move a vehicle in the $x-y$ plane. The outer-loop control law for the leader of the formation, which is characterized by Cartesian state (x_o, y_o, z_o), is given as [14, 15]

$$\begin{aligned} u_{xo} &= k_p(x^* - x_o) - k_d\dot{x}_o, \\ u_{yo} &= k_p(y^* - y_o) - k_d\dot{y}_o, \end{aligned} \tag{3.4}$$

where x^* and y^* stand for the reference trajectory to be followed by the leader, and x_o, y_o are positions in inertial frame Cartesian coordinates of the leader vehicle. Gain K_i is equal to zero.

3.1.2 *Quadrotor dynamics and control*

Typical quadrotor UAVs can be represented schematically as in Fig. 3.8. Four propellers generate thrust. The propellers can be commanded such that the vehicle exhibits certain attitudes and tracks required trajectories. Simplified dynamics of the ith quadrotor vehicle, obtained with Euler-Lagrange equations in closed loop with an inner-loop control law, are given as follows [10]:

$$\begin{aligned} m\frac{d^2x_i}{dt^2} &= -u_i \sin\theta_i \\ m\frac{d^2y_i}{dt^2} &= u_i \cos\theta_i \sin\phi_i \\ m\frac{d^2z_i}{dt^2} &= u_i \cos\theta_i \cos\phi_i - mg \\ \frac{d^2\psi_i}{dt^2} &= \tau_{\psi,i} \\ \frac{d^2\theta_i}{dt^2} &= \tau_{\theta,i} \\ \frac{d^2\phi_i}{dt^2} &= \tau_{\phi,i} \end{aligned} \tag{3.5}$$

where m is the mass of the vehicle, x_i and y_i are the coordinates of the center of mass in the fixed $x - y$ plane, z_i is the altitude, ψ_i is the yaw

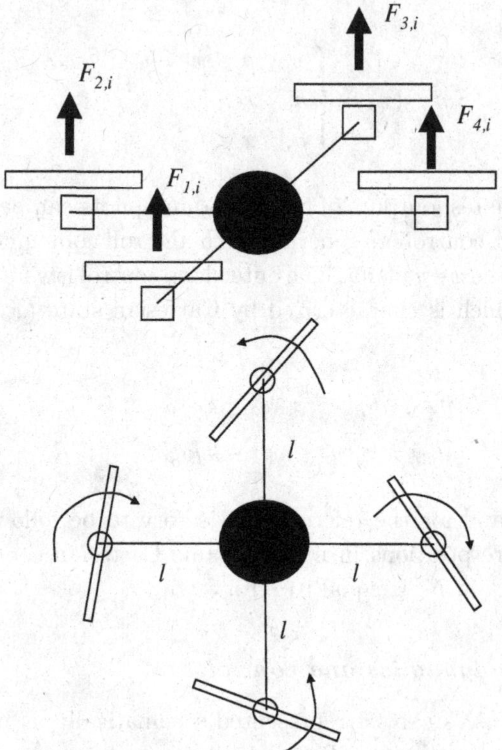

Fig. 3.8 Two views of quadrotor aircraft with four symmetric propellers generating forces $F_{1,i}$ to $F_{4,i}$.

angle, θ_i is the pitch angle, and ϕ_i is the roll angle. Control inputs are given as

$$\begin{aligned} u_i &= F_{1,i} + F_{2,i} + F_{3,i} + F_{4,i} \\ \tau_{\psi,i} &= \tau_{M_{1,i}} + \tau_{M_{2,i}} + \tau_{M_{3,i}} + \tau_{M_{4,i}} \\ \tau_{\theta,i} &= (F_{2,i} - F_{4,i}) \cdot l \\ \tau_{\phi,i} &= (F_{3,i} - F_{1,i}) \cdot l \end{aligned} \tag{3.6}$$

where forces $F_{1,i}$ to $F_{4,i}$ depend on the speed of the motors, $\tau_{M_{j,i}}$ ($j = 1, 2, 3, 4$) is the couple produced by motor $M_{j,i}$, and l is the shortest distance between each motor and the center of mass.

Plant and low-level stabilizing controller parameters can be found in Tables 3.2 and 3.3 in Ref. [10]. Trajectory tracking controllers for the $x - y$

plane motion of a quadrotor are given as follows:

$$u_i = -a_{z1}\dot{z}_i - a_{z2}(z_i - z_d), \tag{3.7}$$

$$\begin{aligned}\tau_{\phi,i} = &-\sigma_{\phi1}(\dot{\phi}_i + \sigma_{\phi2}(\phi_i + \dot{\phi}_i + \sigma_{\phi3}(2\phi_i + \dot{\phi}_i + \tfrac{1}{g}\dot{y}_i \\ &+\sigma_{\phi4}(3\phi_i + \dot{\phi}_i + \tfrac{3}{g}\dot{y}_i + \tfrac{k_p\cdot(y_i-y_d)+\int k_i^c\cdot(y_i-y_d)dt+k_d\cdot\dot{y}_i}{g})))),\end{aligned} \tag{3.8}$$

$$\begin{aligned}\tau_{\theta,i} = &-\sigma_{\theta1}(\dot{\theta}_i + \sigma_{\theta2}(\theta_i + \dot{\theta}_i + \sigma_{\theta3}(2\theta_i + \dot{\theta}_i - \tfrac{1}{g}\dot{x}_i \\ &+\sigma_{\theta4}(3\theta_i + \dot{\theta}_i - \tfrac{3}{g}\dot{x}_i + \tfrac{k_p\cdot(x_i-x_d)+\int k_i^c\cdot(x_i-x_d)dt+k_d\cdot\dot{x}_i}{g})))).\end{aligned} \tag{3.9}$$

Controller parameters k_p, k_d, k_i^c, a_{z1}, and a_{z2} can be tuned by means of simulations. Saturation functions $\sigma_{\phi1}$, $\sigma_{\phi2}$, $\sigma_{\phi3}$, $\sigma_{\phi4}$, $\sigma_{\theta1}$, $\sigma_{\theta2}$, $\sigma_{\theta3}$, and $\sigma_{\theta4}$ are bounded from above. The values of the various parameters and the simulation results are given in Sec. 3.6.2. Equations (3.8) and (3.9) are PID variants of the controllers proposed in Ref. [10]. Control inputs (3.8) and (3.9) provide tracking of (x_d, y_d) on the $x-y$ plane for the simplified quadrotor model (3.5).

3.1.3 *Actuator faults*

In this subsection, we define actuator faults as being either faults on control surfaces or on motors driving the control surfaces. Two categories of faults are presented: abrupt and non-abrupt faults. To define the so-called actuator faults, let us start by considering the signal of interest, which we denote as $F_i(t)$. This signal can represent either an actuator output signal $F_{ji}(t)$ in Eqs. (3.1), a control input signal, or an exogenous force, as suggested in Eqs. (3.1) and (3.2), for the ALTAV control system. For the quadrotor, $F_i(t)$ corresponds to the control inputs of Eq. (3.6). Let t_f denote the time at which a component-level fault actually occurs. $F_i(t)$ can be formally expressed, for vehicle $i \in \mathcal{V}$, as $F_i(t) = K_i(t_f, t, F_{in}(t))$, where $F_{in}(t)$ denotes the nominal signal, exempt from fault. K_i is the identity function, i.e. $F_i(t) = F_{in}(t)$, when $t < t_f$ and a polynomial function of F_{in} when $t \geq t_f$. Identity function K_i is typically non-differentiable and possibly discontinuous at t_f. Let $F_i(t)$ be expressed as

$$\begin{aligned}F_i(t) &= F_{in}(t) - F_{in}(t) + K_i(t_f, t, F_{in}(t)) \\ &= F_{in}(t) + 1_{t-t_f}(K_i(t_f, t, F_{in}(t)) - F_{in}(t)),\end{aligned} \tag{3.10}$$

where 1_{t-t_f} is the unit step function, which is equal to 1 if $t \geq t_f$, and 0 otherwise. Vehicle dynamics can be modeled as a closed-loop nominal system excited by disturbance

$$1_{t-t_f}(K_i(t_f, t, F_{in}(t)) - F_{in}(t)).$$

An uncompensated fault can be represented by means of an additive disturbance δ_i on the right-hand side of Eqs. (3.1), for the ALTAV, and on the right-hand side of Eq. (3.6), for the quadrotor. Faulty signal δ_i belongs to the following set

$$\begin{aligned}\Omega_\delta = \{&\delta_i \in \mathbb{R}^3\backslash\{0,0,0\}; i \in \mathcal{V}\backslash\{0\}, \delta_i = 1_{t-t_f} d_i(t),\\ &\|d_i(t)\| \geq D_i > 0 \text{ for all } t \geq t_f\}.\end{aligned} \tag{3.11}$$

The representation of a fault by the addition of a disturbance can be applied to general affine-in-the-input state-space models such as the one used in Sec. 3.3 for the design of the decentralized fault detector. The definition of Ω_δ in Eq. (3.11) pertains to faults characterized by an abrupt jump, which comes from the use of the unit step function 1_{t-t_f}. Faults whose dynamics are much faster than those of the nominal vehicle can be interpreted as abrupt [92]. Abrupt changes are modeled by means of discontinuous functions $\delta_i(t_f)$, where $\delta_i(t_f^+) - \delta_i(t_f^-)$ is sufficiently large to yield fast time-drifting response of the closed-loop dynamics. In Eq. (3.11), entries of d_i are polynomial functions of time and $D_i > 0$ is a lower bound on the Euclidean norm of signal d_i, such that all faults $\delta \in \Omega_\delta$ are detectable. Conditions about D_i are derived in the next section. Exogenous disturbance δ_i can originate from either an exogenously or endogenously generated fault or failure. For instance, F_{in} involved in an actuator fault represents the control signal and thus makes δ_i endogenous to the closed-loop system, although δ_i is not expressed analytically. Disturbance δ_i may be triggered by an event that is external to the system, such as a projectile penetrating the body of the vehicle.

Typical actuator faults are shown in Fig. 3.9. Faults include lock in place (LIP), hard-over fault (HOF), float and loss of effectiveness (LOE) [93]. Body damage of the ALTAV may result in a loss of buoyancy (LOB). LOB can either be sudden, with the time to reach approximately zero buoyancy being relatively close to zero, $t_{F_b} \approx 0$, or gradual ($t_{F_b} > 0$), as shown in Fig. 3.9(d). Two types of abrupt faults can typically occur with the ALTAV: HOF with $1 - u(t_f)/u_{HOF} \approx 1$ and $\theta_{HOF} \approx \pi/2$, and sudden loss of buoyancy. For the quadrotor, HOF is an abrupt fault. Particular ALTAV and quadrotor actuator faults of types LIP and float, which are characterized, for $t \geq t_f$, by

$$\delta_i(t) = \begin{cases} u(t_f) & \text{(LIP)}, \\ 0 & \text{when } u(t_f) = 0 \text{ (Float)}, \end{cases} \tag{3.12}$$

as well as HOF with $\theta_{HOF} \ll \pi/2$ and gradual LOB, strictly for the ALTAV, constitute the so-called class of non-abrupt faults considered in this book.

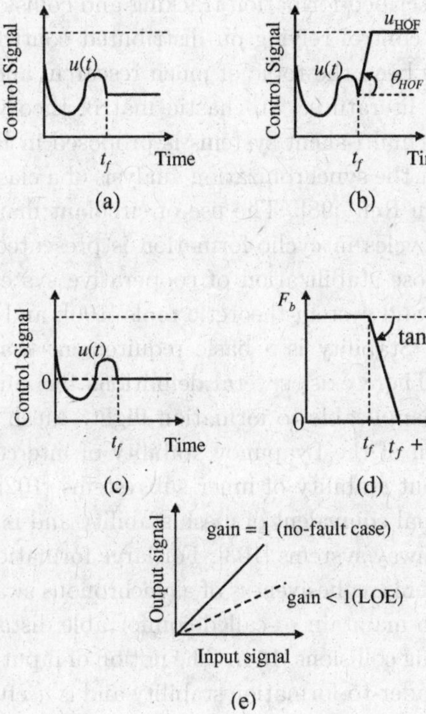

Fig. 3.9 Actuator faults of type: (a) lock in place, (b) hard over, (c) float, (d) loss of buoyancy, and (e) loss of effectiveness.

Briefly, non-abrupt faults are characterized by slower dynamics than that found with abrupt faults. Techniques used to detect the occurrence of faults may be better suited for one type of fault than another. This will be further discussed in Secs. 3.3 and 3.4.

3.2 Formation Control

A great number of techniques for formation flight control are available in the literature. String stability of leaderless vehicle platoons with ring coupling is proposed in Ref. [94] by means of identical controllers and circulant matrix arguments. Cooperative control of nonholonomic agents for target tracking and formation control is investigated in Ref. [95], where communication delays are taken into account. Decentralized potential-function-based smooth control laws are designed in Ref. [96] to allow unicycle-type mobile robots to achieve prescribed formation tracking and collision avoidance objectives. Formation control relying on distributed controllers following neighboring rules has been the topic of much research, and several results can be found in the literature. Stochastic matrix theory for the analysis of jointly connected multi-agent systems is proposed in Ref. [97]. Partial contraction theory for the synchronization analysis of a class of coupled networks is developed in Ref. [98]. The use of circulant matrices for the analysis of multiple unicycles in cyclic formation is presented in Ref. [99]. Some techniques propose stabilization of cooperative systems from set-valued Lyapunov theory and system-theoretic tools [100], and classical Lyapunov techniques [101]. Stability is a basic requirement that a control system should satisfy. There exist several definitions for multi-agent systems stability, which are applicable to formation flight. Mesh stability is defined as the combination of the Lyapunov stability of interconnected systems with the input-output stability of inner subsystems [102]. String stability is the one-dimensional equivalent of mesh stability and is of interest mainly in automated highway systems [103]. For large formations, such as swarms of unmanned systems, cohesiveness of asynchronous swarms has been defined as the ability to maintain so-called comfortable distances between neighbors while avoiding collisions [104]. The notion of input-to-state stability forms the basis of leader-to-formation stability and is useful for the study of formation sensitivity to interconnection topology, as presented in Ref. [105].

In this book, we consider a UAV formation in which the line-of-sight (LOS) angle of each pair of vehicles and inter-vehicle distances are required to remain constant with respect to the bodyframe of the leader. It is assumed that the leader tracks a predetermined trajectory whereas the followers seek to maintain local relative positions and speeds. Elements of contraction theory [106] are used to prove the exponential convergence of the trajectory of each unicycle to a ball centered on a prescribed trajectory. The latter depends on the formation geometry commands and on the position of the leader. The contraction property of the error dynamics is obtained provided the symmetric part of its Jacobian is uniformly negative definite. The control law achieves contraction of a UAV formation in three steps. First, formation dynamics are written as a subsystem feedback decomposition. Second, a set of feasible linear matrix inequalities (LMIs) is derived such that the feedback subsystem characterizing unicycle interconnection is strictly positive real (SPR). Third, the whole formation is proved to be contracting provided a set of bilinear matrix inequalities (BMIs) is satisfied. More details on LMIs and BMIs are given in Refs. [107–109]. The distributed formation control law presented in this section ensures convergence of the trajectories of the unicycle models to a prescribed ball, robust with respect to exogenous disturbances such as wind turbulence. The formation control scheme is a constructive design approach, enabling controller gain tuning and multi-objective control problem solving by means of the well-known LMI formalism, although at the cost of potentially conservative performances.

A point must be made regarding the formations considered in this book. Leader-to-follower formations are weakened by a single point of failure, the leader, by definition. The leader may experience problems during mission, such as a system malfunction or a crash caused by hostile fire or adverse weather. The formation can be made robust to the failure of one or more of its members. For example, by assigning a virtual leader to a formation, in which case each vehicle relies on a local controller to maintain the required separation from the virtual entity while the formation carries out consensus building to ensure effective performance as noise, disturbances and changes in the environment affect operations [110]. Another approach allows for several leaders in a given formation; namely, those having the required onboard components to assume the role of leader in case the current leader experiences problems. Thus, to ensure mission reliability, the followers must be able to detect if the current leader should be replaced and, if so, by which one of them. For a team with a single, predetermined leader, the latter may be equipped with additional safety features and capabilities, as a precautionary measure, and be operating remotely from zones of risk, away from the rest of the fleet. In such case, communications become even more critical to ensure mission success, and the concept of a tight formation no longer applies. To limit the scope of the book and to present concepts of safety and reliability in cooperating UAVs in a uniform manner, the health management functions are presented for formations of homogeneous leader-to-follower vehicles. It must be stressed that the functionalities discussed in this book can, however, be extended to other types of formations.

3.2.1 *Elements of contraction theory*

The leader-to-follower formation control law presented in this section relies on contraction theory. Time-domain analysis of the behavior of dynamic systems is often characterized by the convergence of the trajectory to an attractor and is typically achieved by means of Lyapunov techniques.

Lohmiller and Slotine recently introduced the contraction theory in Ref. [106], where the exponential convergence in time of trajectories of smooth nonlinear systems expressed as

$$\dot{x} = f(x,t) \tag{3.13}$$

is analyzed by using virtual displacements

$$\delta z = \Omega(x,t)\delta x,$$

where $\Omega(x,t)$ is a uniformly invertible square matrix. For any fixed time instant t, δz and δx are infinitesimal displacements. Let us define the symmetric and uniformly positive definite matrix

$$P = \Omega^T \Omega.$$

Exponential convergence to zero, over time, of δz, and thus of δx, is obtained if the time derivative of the squared length $\delta z^T \delta z$, expressed as

$$d(\delta z^T \delta z)/dt = 2\delta z^T F \delta z,$$

is such that the symmetric part of the generalized Jacobian

$$F = \left(\dot{\Omega} + \Omega \frac{\partial f}{\partial x}\right) \Omega^{-1} \tag{3.14}$$

is uniformly negative definite in a region called the contraction region. From Theorem 2 in Ref. [106], any trajectory of Eq. (3.13) starting in a ball of constant radius with respect to P, centered about a given trajectory and contained at all times in a contraction region remains in the ball and converges exponentially to this trajectory. In the sequel, a constant matrix Ω is used along with the fact that if Eq. (3.13) is disturbed by $d(x,t)$, i.e.

$$\dot{x} = f(x,t) + d(x,t), \tag{3.15}$$

where $\|d\|$ is bounded, then the distance R defined as the smallest path integral

$$R = \int_{P_1}^{P_2} \|\delta x^T P \delta x\|$$

between trajectories P_1 of Eq. (3.13) and P_2 of Eq. (3.15), verifies [106]

$$\dot{R} + |\lambda_{\max}|\, R \leq \|\Omega d\|\,, \tag{3.16}$$

where $\lambda_{\max}$ denotes the largest eigenvalue of the symmetric part of F and $\|.\|$ stands for the Euclidean norm. It should be pointed out that convergence is understood as a convergence property of trajectories in the time domain.

3.2.2 *Simplified modeling for the purpose of formation control synthesis*

Without loss of generality, to derive the formation control law, each UAV, stabilized by means of appropriately designed autopilots, is modeled as a unicycle, which is composed of two first-order dynamics and kinematic constraints given by Eqs. (3.17). This model presents the salient features of the vehicle dynamics in closed-loop with the autopilots, and provides sufficient details for the design of high-level guidance, navigation and control of UAVs. It is shown in Sec. 3.6 that despite such simplifying modeling assumptions, the formation control and the distributed fault detection schemes presented in this chapter lead to satisfactory closed-loop performances for both ALTAVs and quadrotors. Consider the formation of $n+1$ identical unicycle models comprising a single leader and n followers. The leader corresponds to node 0, whereas the followers constitute nodes 1 to n of a graph

$$\mathcal{G} = (\mathcal{V}, \mathcal{E}),$$

which is defined by a finite set of vertices

$$\mathcal{V} = \{1, ...n\},$$

together with a set of edges given as

$$\mathcal{E} = \{(i,j) : i, j \in \mathcal{V}\}.$$

Any node i in the formation belongs to the set $\mathcal{V}_0$ which is defined as

$$\mathcal{V}_0 = \mathcal{V} \cup \{0\}.$$

The following general rules will be used to define variables: a single subscript is associated to a node; a double subscript is associated to a relative quantity between two nodes; and ˜ identifies an error quantity. The terms unicycle, UAV and node are used as synonyms in this section of the book. The unicycle representing the UAV dynamics is modeled as [111]

$$\begin{aligned} \dot{q}_i &= v_i \begin{bmatrix} \cos\theta_i \\ \sin\theta_i \end{bmatrix}, \\ \dot{v}_i &= \alpha\,(v_i^c - v_i), \\ \dot{\theta}_i &= \beta\,(\theta_i^c - \theta_i), \end{aligned} \tag{3.17}$$

where $i \in \mathcal{V}_0$, $q_i = [x_i \; y_i]^T$ is the planar position coordinates with respect to the center of mass of the vehicle, v_i is velocity magnitude, θ_i is heading angle and control laws are v_i^c and θ_i^c. In Eqs. (3.17), α and β are positive constants of the autopilot. Vehicle positions are obtained with respect

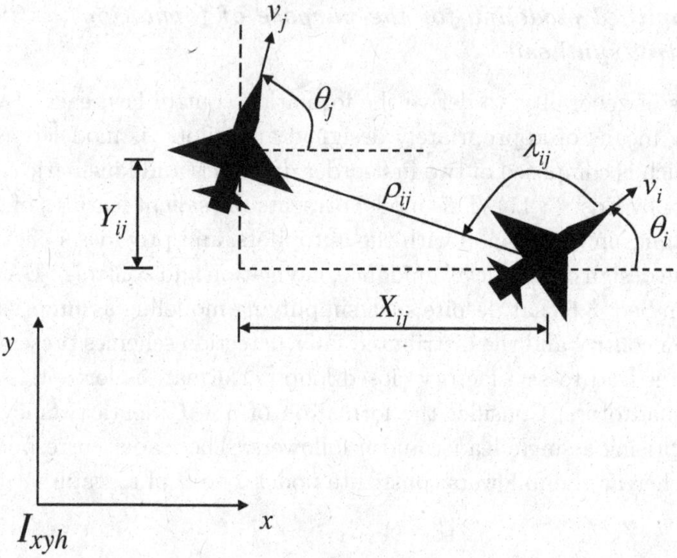

Fig. 3.10 Geometry of neighboring nodes i and j.

to an inertial frame $\mathcal{I}_{xyh}$. Figure 3.10 presents the planar motion of any two neighboring vehicles in the formation. The relative distance between vehicles i and j is denoted as ρ_{ij}, and the LOS angle is λ_{ij}.

The following definition and assumptions characterize the vehicles and the formation.

Definition 3.1. Formation topology. The formation can be characterized as follows.

- Associate to each node i the set $\mathcal{N}_i$ which includes the ith node and all the vehicles j in the formation that are sensed by or that send information to node $i \in \mathcal{V}$. Node i does not send information to $j \in \mathcal{N}_i$. $\mathcal{N}_i$ is thus endowed with a neighboring unidirectional relation ($\rightsquigarrow$); that is, when j is an element of $\mathcal{N}_i$, a directed edge (j, i) exists such that $j \rightsquigarrow i$, which means that information/sensor data, e.g. position and speed about j, flows from j to i. Node j is defined as a neighbor of node i.
- Let $\mathcal{N}$ be the neighboring set defined as

$$\mathcal{N} = \bigcup_{i \in \mathcal{V}} \mathcal{N}_i.$$

The set of nodes $\mathcal{V}$ along with the set of edges

$$\mathcal{E} = \{(i,j) \in \mathcal{V} \times \mathcal{V},\, i \rightsquigarrow j\}$$

constitute a directed acyclic graph (DAG) $\mathcal{G} = (\mathcal{V}, \mathcal{E})$.

- The DAG $\mathcal{G}_0$, which can be expressed as $\mathcal{G}_0 = (\mathcal{V}_0, \mathcal{E}_0)$, comprises the edge set $\mathcal{E}_0$, which is composed of $\mathcal{E}$ and of at least one directed edge $(0, i)$ such that (1) $\mathcal{G}_0$ is always weakly connected, and (2) every node in $\mathcal{V}$ has indegree greater than or equal to 1.
- $\mathcal{I}$ is the set of all vertices i with edges having only the leader as the outgoing vertex. For example, in Fig. 3.11, set $\mathcal{I}$ comprises nodes 1 and 2. The set of terminal vertices is labeled as $\mathcal{O}$. By terminal vertices, we mean vertices whose outdegree is zero, such as nodes 3, 4, and 5 in Fig. 3.11. For each i in $\mathcal{E}$, if

$$\mathcal{N}_i \cap \mathcal{I} \neq \emptyset,$$

let

$$\mathcal{N}_i^* = \mathcal{N}_i \cap \mathcal{I} \tag{3.18}$$

and

$$N_i = \mathcal{N}_i \backslash \mathcal{I}. \tag{3.19}$$

If

$$\mathcal{N}_i \cap \mathcal{I} = \emptyset,$$

this means $N_i = \mathcal{N}_i$. Let

$$n_i = \dim N_i,\ n_i^* = \dim \mathcal{N}_i^*.$$

Finally, the outdegree of node j is labeled as o_j.

- The $n \times n$ identity matrix is denoted as I_n. The $n \times n$ zero matrix is 0_n. The $n \times 1$ vector, whose entries are all unity, is written as 1_n. The conjugate transpose of a complex matrix Z is expressed as Z^H. In this chapter, $|\cdot|$ and $\|\cdot\|$ denote the absolute value and the Euclidean norm, respectively. Finally, $\sum_{j \in \mathcal{N}_i}$ is understood as $\sum_{j \in \mathcal{N}_i, j \neq i}$. A linear normed space of Lebesgue-integrable functions is denoted as L_p for $1 \leq p < \infty$ [112].

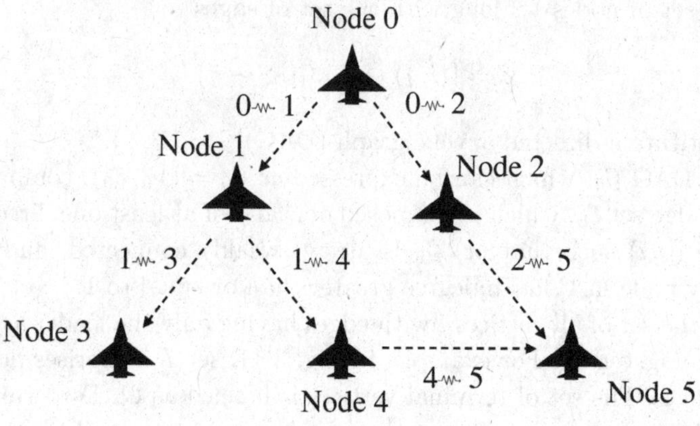

Fig. 3.11 Example of a six-UAV formation. The leader is connected to the zero-indegree nodes of $\mathcal{G}$.

Assumption 3.1. Leader. The single leader vehicle of the formation tracks a reference trajectory $q^* = [x^* \; y^*]^T$ expressed in $\mathcal{I}_{xyh}$. Functions q^*, $\dot{q}^*$ and $\ddot{q}^*$ are available to the leader at all times ($\dddot{q}^* \in L_\infty$). Orientation of the leader is such that $\dot{\theta}_0$, $\ddot{\theta}_0 \in L_\infty$. The trajectory followed by the leader is typically of Dubins type, which can be defined as a series of line segments and circular arcs of constant rate turns [113]. The leader vehicle is capable of communicating with information broadcasting and decision making resources that are exogenous to the formation, and owns a certain level of autonomy in terms of its ability to plan and generate trajectories, and to avoid collisions.

Assumption 3.2. Followers. Signals v_i and θ_i are available to a node i in $\mathcal{V}$. If a node j lies in $\mathcal{N}_i$ then ρ_{ij}, $\dot{\rho}_{ij}$ and the LOS angle λ_{ij} resolved in the bodyframe of the ith node are available to the ith node. These quantities can be projected on $\mathcal{I}_{xyh}$ to obtain

$$q_{ij} = q_i - q_j = \rho_{ij} \begin{bmatrix} \cos(\theta_i + \lambda_{ij}) \\ \sin(\theta_i + \lambda_{ij}) \end{bmatrix}. \tag{3.20}$$

Furthermore, vehicle speed magnitude is bounded as

$$v_m^2 < \left\| \dot{q}_i \right\|^2 < v_M^2 \tag{3.21}$$

which leads to the definition of the following domains:

$$\widetilde{\mathcal{S}}_i = \{\widetilde{q}_i, \dot{\widetilde{q}}_i | \widetilde{q}_i \in \mathbb{R}^2, \left\|\dot{q}_i\right\|^2, \left\|\dot{q}_i^*\right\|^2 \in (v_m, v_M)\}$$
$$\widetilde{\mathcal{D}} = \widetilde{\mathcal{S}_1} \times ... \times \widetilde{\mathcal{S}_n},$$

where $\widetilde{q}_i = q_i - q_i^*$, and $\dot{\widetilde{q}}_i = \dot{q}_i - \dot{q}_i^*$. Signal q_i^* is introduced in Assumption 3.3.

Assumption 3.3. Formation geometry. The required geometry of the formation is determined by (1) a reference trajectory being tracked by the leader (Assumption 3.1), and (2) required relative positions of the nodes expressed in the inertial frame $\mathcal{I}_{xyh}$ and defined as

$$\begin{aligned} q_{ij}^* &= q_i^* - q_j^* \\ &= \rho_{ij}^* \begin{bmatrix} \cos(\theta_0 + \lambda_{ij}^*) \\ \sin(\theta_0 + \lambda_{ij}^*) \end{bmatrix}, \end{aligned} \tag{3.22}$$

where ρ_{ij}^* and λ_{ij}^* denote, respectively, the prescribed relative distance and LOS angle between i and j of $\mathcal{V}_0$. In Eq. (3.22), q_i^* is the required planar position of the ith node; however, this position is not explicitly used by the formation control law. The formation geometry is thus determined by two time-invariant parameters, ρ_{ij}^* and λ_{ij}^*, and by the fact that, at equilibrium, (1) each vehicle keeps a heading angle equal to that of the leader (θ_0), and (2) λ_{ij}^* remains unchanged in the bodyframe of the leader.

Assumption 3.4. Exogenous disturbance. The tracking error of the leader is given by $\widetilde{q}_0 = q_0 - q^*$, and is possibly corrupted by an additive exogenous disturbance δ_0, which is in $L_2 \cap L_\infty$. The derivative of δ_0 with respect to time is assumed to lie in L_∞. Such disturbance may correspond to measurement noise or abrupt changes in reference trajectories. Wind turbulence is of interest, especially when considering MAVs and SUAVs. The Dryden model is used for such purpose in Sec. 3.6.

From the modeling assumptions, we can make the following statements prior to stating the formation control objective. The formation is such that the leader is connected to at least one node of $\mathcal{G}$. DAG $\mathcal{G}$ is not required to be connected, although $\mathcal{G}_0$ must be weakly connected to ensure that the formation is stabilizable. Expressing the relative positions and velocities with respect to inertial frame $\mathcal{I}_{xyh}$ simplifies somewhat the stability analysis of the formation and allows expressing the complete formation geometry in the same referential axes. The knowledge of q^*, and q^*_{ij}, for $i \in \mathcal{V}$ and $j \in \mathcal{N}_i$, as stated in Assumptions 3.1 and 3.3, is sufficient to determine the required position q^*_i for each node i such that each vehicle of the formation is characterized by the orientation angle θ_0 and relative quantities ρ^*_{ij} and λ^*_{ij}. Since the graph is weakly connected, it is possible to determine ρ^*_{i0} and λ^*_{i0} such that the position of node i depends only on θ_0 and q^*; that is,

$$q^*_i = q^* + \rho^*_{i0} \begin{bmatrix} \cos(\theta_0 + \lambda^*_{i0}) \\ \sin(\theta_0 + \lambda^*_{i0}) \end{bmatrix}, \tag{3.23}$$

where ρ^*_{i0} and λ^*_{i0} are functions of ρ^*_{kl} and λ^*_{kl} for all k and l in $\mathcal{V}_0$ such that the set of undirected edges (k, l) in $\mathcal{E}_0$ constitutes the path that links node i to leader 0.

3.2.3 *Formation control objective*

A distributed control law v^c_i, θ^c_i applied to the ith follower unicycle with model given in Eqs. (3.17) should be such that the tracking error $\widetilde{q}_i$ converges exponentially to a ball centered on $(0, 0)$ with a radius that is an increasing function of $\left|\dot{\theta}_0\right|$, $\left|\ddot{\theta}_0\right|$, $\left\|\ddot{q}^*\right\|$ and $\|\delta_0\|$. Without loss of generality, the exogenous disturbance is assumed to affect only the leader and is labeled as δ_0. The reader is referred to Ref. [114] for the case of unicycle-type followers which are also subject to exogenous disturbances. The formation control scheme should allow for changes in geometries. Recall that formation geometry commands are transmitted from one vehicle to another by means of the communication network. The operating crew waits for a formation to stabilize around a prescribed geometry before commanding a geometry change.

To satisfy the formation control objective, a two-loop distributed control law is presented [115]. The inner loop consists of a feedback linearization scheme $v^c_i(v_i, \theta_i, a_i)$ and $\theta^c_i(v_i, \theta_i, a_i)$ applied to the speed and heading angle control channels, respectively. Feedback linearization is performed on a stabilized UAV modeled as a unicycle. The outer-loop control law aims at contributing to the formation stabilization by using local information; that is, relative quantities with respect to neighbors. The outer loop in a_i achieves uniform positive negativeness of the Jacobian associated with the first variation of the error dynamics of the stabilized formation by building upon contraction theory, as discussed in Sec. 3.2.1. The first variation is interpreted as a feedback subsystem decomposition. More precisely, the feedforward subsystem is represented as a quasi-linear model

$$\dot{\delta x} = A(x)\delta x + B\delta u,$$

where $A(x)$ lies within a fixed polytope of matrices as long as state x remains in a prescribed domain. The feedback subsystem, which represents the interconnection grid, is characterized by a constant matrix $H+\Gamma$ that is made strictly positive real, or SPR. Then, with states starting in the prescribed domain $\widetilde{\mathcal{D}}$, contraction of the closed-loop system is obtained provided a set of matrix inequalities is satisfied at the vertices of the polytope.

3.2.4 *Closed-loop representation of the formation*

The formation, as a closed-loop system, is under feedback linearizing [33] and passivating [116, 117] control, as schematically illustrated in Fig. 3.12. Feedback linearization is the process by which a nonlinear system is made linear by means of feedback control. Passivating a system amounts to designing a control law that renders a system passive for a given input-output signal pair. In the figure, the internal feedback loops are illustrated for the ith UAV, although this block diagram is the same for all follower

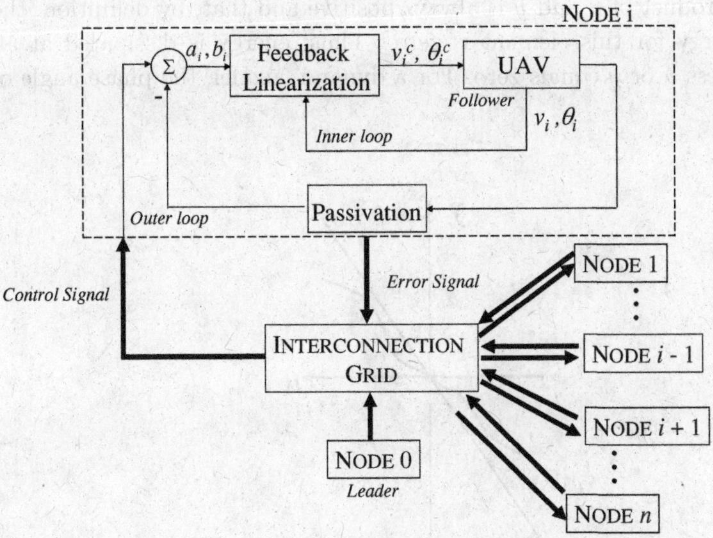

Fig. 3.12 Block diagram of inner- and outer-loop controllers for formation flying UAVs (nodes).

UAVs of the formation, namely nodes 1 to n. The interconnection grid represents the dynamic interactions among the vehicles of the formation. These interactions are determined by the outer-loop control law, which includes the passivating loop.

Recall that for multi-input, multi-output nonlinear systems, passivity expresses an energy-like balance for input-output pairs characterized by [33]

$$\underbrace{\int_{t_0}^{t} u^T(\tau)y(\tau)d\tau}_{\text{Supplied energy}} = \underbrace{S(t) - S(t_0)}_{\text{Stored energy}} + \underbrace{\delta \int_{t_0}^{t} \|y(\tau)\|^2 \, d\tau + \varepsilon \int_{t_0}^{t} \|u(\tau)\|^2 \, d\tau}_{\text{Dissipated energy}}. \tag{3.24}$$

In Eq. (3.24), if δ and ε are zero, the system is lossless. If $\delta > 0$ (respectively, $\varepsilon > 0$), the system is strictly output passive (respectively, strictly input passive); that is, dissipation occurs at the output or the input, or both. In other words, a passive system is a system that cannot store more energy than supplied. For example, effects of exogenous disturbances in L_2 or in L_∞ can be attenuated by using the concept of passivity. A memoryless nonlinearity restricted to the first and third quadrants, as illustrated in Fig. 3.13, is passive if the u-axis is included in the function definition space, and strictly passive otherwise. This can be shown by using Eq. (3.24), given that the product of u and y is always positive and that, by definition, the stored energy for this element is zero. Thus, energy is dissipated at all times, unless u or y equals zero. For a dynamic model, the phase angle of

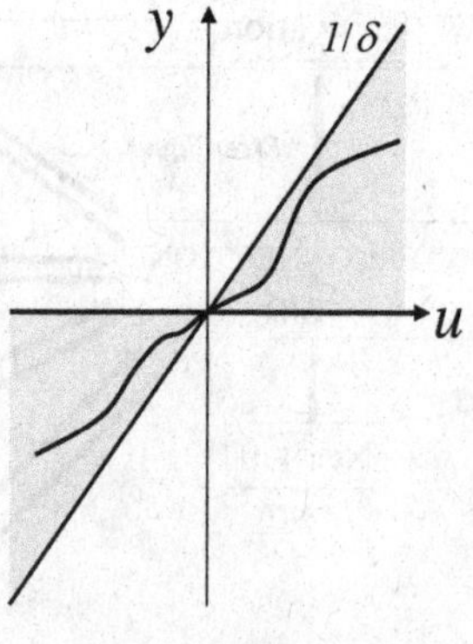

Fig. 3.13 Input-output sector of passive systems.

passive (respectively, strictly passive) linear systems is within $[-\pi/2, \pi/2]$ radians (respectively, $(-\pi/2, \pi/2)$ radians).

Passivity is closely related to the property of positive realness. Referring to Eq. (3.24), a system is positive real if

$$\int_0^{+\infty} u^T(\tau)y(\tau)d\tau \geq 0.$$

Passivity and positive realness are equivalent concepts in the case of causal systems [112]. The reader is referred to Ref. [118] for a complete set of positive realness conditions. Strict positive realness of a system H is obtained when $H - \varepsilon$ is positive real for some $\varepsilon > 0$ [112]. Some passivity properties are applicable to closed-loop, interconnected systems, and thus apply to the case of formation control. Indeed, the passivity theorem states that a system obtained by connecting two passive systems in a negative feedback loop is passive. Furthermore, bounded-input, bounded-output stability is obtained if the forward path of the passive closed-loop system is strictly output passive [33].

The control laws in $v_i^c(v_i, \theta_i, a_i)$, $\theta_i^c(v_i, \theta_i, a_i)$ are in $\mathbb{R}$, whereas a_i is in $\mathbb{R}^2$. Such control laws are used to perform the feedback subsystem decomposition, and hence the constructive stability analysis. The latter being in the sense of convergence of state trajectories. The control laws are defined as follows.

Definition 3.2. Inner- and outer-loop control laws. Consider C_i in $\mathbb{R}^{2\times 4}$ and symmetric SPR matrices H_i, K_i in $\mathbb{R}^{2\times 2}$. Define

$$\Gamma_i = \gamma_i \left(K_i + \frac{I_2}{n_i}\right)$$

where $\gamma_i = 1$ if node i is connected to the leader 0 (if $0 \in \mathcal{N}_i$); otherwise $\gamma_i = 0$. Then, the control law applied to each node i in $\mathcal{V}_0$ is defined as the series combination of the inner-loop control law expressed as

$$\begin{bmatrix} v_i^c \\ \theta_i^c \end{bmatrix} = \begin{bmatrix} v_i \\ \theta_i \end{bmatrix} + \begin{bmatrix} \frac{\cos\theta_i}{\alpha} & \frac{\sin\theta_i}{\alpha} \\ -\frac{\sin\theta_i}{\beta v_i} & \frac{\cos\theta_i}{\beta v_i} \end{bmatrix} a_i, \tag{3.25}$$

with the outer-loop control law expressed, for all i in $\mathcal{V}$, as

$$a_i = -\sum_{j\in\mathcal{N}_i} (H_i + \frac{I_2}{n_i})C_i(\mathcal{Q}_i - \mathcal{Q}_j - \mathcal{Q}_{ij}^o) - \Gamma_i(C_i(\mathcal{Q}_i - \mathcal{Q}_{i0}^o) - Z_0) \tag{3.26}$$

where

$$\mathcal{Q}_i = \begin{bmatrix} q_i \\ \dot{q}_i \end{bmatrix},$$

$$Z_0 = C_0 \mathcal{Q}_0 + \delta_0,$$

$$\mathcal{Q}_{ij}^o = \begin{bmatrix} q_{ij}^{oT} \\ 0_{21} \end{bmatrix},$$

and

$$q_{ij}^o = \rho_{ij}^* \left[\cos(\theta_i + \lambda_{ij}^*) \ \sin(\theta_i + \lambda_{ij}^*) \right]. \tag{3.27}$$

The outer-loop control law applied to the leader ($i = 0$) is given as

$$a_0 = -C_0 \widetilde{Z}_0^m + \ddot{q}^*, \tag{3.28}$$

where C_0 is in $\mathbb{R}^{2\times 4}$, $\widetilde{Z}_0^m = \widetilde{\mathcal{Q}}_0 + \delta_0$ and $\widetilde{\mathcal{Q}}_0 = \mathcal{Q}_0 - \mathcal{Q}_0^*$.

The outer-loop control law (3.26) requires the knowledge of the relative position and velocity of node i with respect to its connected neighbors, which belong to $\mathcal{N}_i$. In Eq. (3.25), velocity v_i and constants α and β are assumed to be nonzero for all time instants. Under the action of the inner-loop control law (3.25), the linearized unicycle model

$$\ddot{q}_i = a_i \tag{3.29}$$

in closed loop with the outer-loop control law (3.26) yields

$$\begin{aligned} \dot{\widetilde{\mathcal{Q}}}_i &= A_i \widetilde{\mathcal{Q}}_i + B_i w_i \\ \widetilde{Z}_i &= C_i \widetilde{\mathcal{Q}}_i, \theta_i = \Theta(\widetilde{\mathcal{Q}}_i + \mathcal{Q}_i^*) \end{aligned} \tag{3.30}$$

where

$$\widetilde{\mathcal{Q}}_i = \mathcal{Q}_i - \mathcal{Q}_i^*,$$

$$A_i = A^{aux} - B_i C_i,$$

$$A^{aux} = \begin{bmatrix} 0_2 & I_2 \\ 0_2 & 0_2 \end{bmatrix},$$

$$B_i = \begin{bmatrix} 0_2 \\ I_2 \end{bmatrix},$$

and

$$\begin{aligned} w_i &= -r_i - u_i + \ddot{q}_i^* + \Gamma_i \widetilde{Z}_0, \\ r_i &= \textstyle\sum_{j \in \mathcal{N}_i} \left(H_i \left(\widetilde{Z}_i - \widetilde{Z}_j \right) - \frac{I_2}{n_i} \widetilde{Z}_j \right) + \gamma_i K_i \widetilde{Z}_i, \\ u_i &= \Delta_i(\theta_i) \\ &= \textstyle\sum_{j \in \mathcal{N}_i} \left(H_i + \frac{I_2}{n_i} \right) C_i \left(\mathcal{Q}_{ij}^* - \mathcal{Q}_{ij}^o \right) - \Gamma_i C_i \left(\mathcal{Q}_{i0}^* - \mathcal{Q}_{i0}^o \right). \end{aligned} \tag{3.31}$$

Function Θ in Eqs. (3.30) depends on state $\widetilde{\mathcal{Q}_i}$ by noticing that

$$\theta_i = \tan^{-1}(y_i/x_i).$$

Nonlinear function Δ_i depends on θ_i, from q_{ij}^o in Eq. (3.27). Formation dynamics are described by Eqs. (3.30) and (3.31), for all i belonging to $\mathcal{V}$, and can be represented by the closed-loop system shown in Fig. 3.14, where

$$A_{fw} = \underbrace{\begin{bmatrix} A_1 & 0 & \cdots & 0 \\ 0 & A_2 & \ddots & \vdots \\ \vdots & \ddots & \ddots & 0 \\ 0 & \cdots & 0 & A_n \end{bmatrix}}_{=diag(A_1,\dots,A_n)},$$

$$\begin{aligned} B_{fw} &= diag(B_1, ..., B_n), \\ C_{fw} &= diag(C_1, ..., C_n), \\ \Gamma &= [\Gamma_1^T ... \Gamma_n^T]^T, \\ \Delta &= diag(\Delta_1, ..., \Delta_n), \\ V &= [w_1 ... w_n]^T, \\ I_{H+\Gamma} &= [r_1 ... r_n]^T, \\ \widetilde{Z} &= [\widetilde{Z}_1^T ... \widetilde{Z}_n^T]^T, \\ U &= [u_1 ... u_n]^T. \end{aligned}$$

Output $I_{H+\Gamma}$ of linear map $H + \Gamma$ is in feedback with Eqs. (3.30). The linear map $H + \Gamma$ is a matrix representing the interconnection among the vehicles, shown in Fig. 3.12 and detailed in the proof of Theorem 3.1. In case of the outer-loop control law derived in Ref. [114], which ensures stability of the formation in translational motion, Δ is zero. This is so since $q_{ij}^o = q_{ij}^*$; that is, q_{ij}^o in Eq. (3.27) becomes

$$q_{ij}^o = \rho_{ij}^* [\cos \lambda_{ij}^* \;\; \sin \lambda_{ij}^*].$$

The closed-loop system is excited by the commanded acceleration $\ddot{q}_i^*$ of vehicle i in $\mathcal{V}$, by the heading angle of the leader, θ_0, and by the tracking

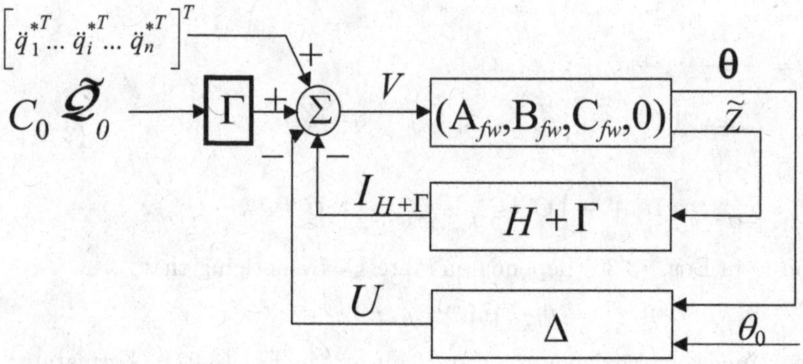

Fig. 3.14 Feedback system interpretation of formation control system.

error of the leader, given as $C_0\widetilde{\mathcal{Q}}_0 \in L_2 \cap L_\infty$, which is asymptotically stable since (3.28) in closed loop with (3.29), with i set to zero, results in

$$\dot{\widetilde{\mathcal{Q}}}_0 = A_0\widetilde{\mathcal{Q}}_0 - B_0C_0\delta_0. \tag{3.32}$$

In Eq. (3.32), A_0, expressed as $A_0 = A^{aux} - B_0C_0$, is Hurwitz. Disturbance δ_0 is defined in Assumption 3.4.

3.2.5 *Convergence analysis of state trajectories*

Knowing the equations and the block diagram for the formation control system, we want to analyze the behavior of the formation under various conditions. More precisely, we would like to predict the behavior of the formation under certain excitations. Recall that the formation of unicycle models of UAVs is under nominal, non-degraded operation. We seek conditions for which the formation control law guarantees stability despite the presence of exogenous disturbances affecting the fleet. Stability of the formation is in the sense of convergence of state trajectories. The main result is given in the form of Theorem 3.1. However, some development is needed prior to presenting the theorem. Let us write the error dynamics in (3.30), for all $i \in \mathcal{V}$, as

$$\dot{\mathrm{X}} = \underbrace{A_{fw}\mathrm{X} + B_{fw}\left(-(H+\Gamma)\,\widetilde{Z} - \Delta_1(\theta_1, ..., \theta_n)\right)}_{=f(\mathrm{X})} + \underbrace{B_{fw}\left(-\Delta_2 + \Gamma\widetilde{Z}_0\right)}_{=d(\theta_0,\dot{\theta}_0,\ddot{\theta}_0,\ddot{q}^*)}. \tag{3.33}$$

Signal $d(\theta_0, \dot{\theta}_0, \ddot{\theta}_0, \ddot{q}^*)$ reflects the impact of the trajectory of the leader on the error dynamics of the follower vehicles. The error signal is defined as $\mathrm{X} = [\widetilde{\mathcal{Q}}_1^T \ ... \ \widetilde{\mathcal{Q}}_n^T]^T$. Also, let us define the following:

$$\Delta_1(\theta_1, ..., \theta_n) = diag(\Delta_{1,1}(\theta_1), ..., \Delta_{n,1}(\theta_n))$$
$$\Delta_2 = diag(\Delta_{1,2}, ..., \Delta_{n,2})$$

whose entries are computed from u_i, and $\ddot{q}^*$ in Eqs. (3.30), for all $i \in \mathcal{V}$, as follows

$$\begin{aligned} \Delta_{i,1}(\theta_i) &= \textstyle\sum_{j \in \mathcal{N}_i} \left(H_i + \frac{I_2}{n_i}\right) C_i \begin{bmatrix} q_{ij}^* - q_{ij}^o \\ 0_{21} \end{bmatrix} + \Gamma_i C_i \begin{bmatrix} q_{i0}^* - q_{i0}^o \\ 0_{21} \end{bmatrix}, \\ \Delta_{i,2} &= -\ddot{q}_i^* + \textstyle\sum_{j \in \mathcal{N}_i} \left(H_i + \frac{I_2}{n_i}\right) C_i \begin{bmatrix} 0_{21} \\ \dot{q}_{ij}^* \end{bmatrix} + \Gamma_i C_i \begin{bmatrix} 0_{21} \\ \dot{q}_{i0}^* \end{bmatrix}. \end{aligned} \tag{3.34}$$

Note that Δ_1 depends on the actual angles $(\theta_1, ..., \theta_n)$ of the unicycle models, while Δ_2 depends on θ_0, $\dot{\theta}_0$, $\ddot{\theta}_0$, and $\ddot{q}^*$ since, from (3.23), acceleration $\ddot{q}_i^*$ in (3.34) can be written as

$$\ddot{q}_i^* = \ddot{q}^* - \rho_{i0}^* \begin{bmatrix} \dot{\theta}_0^2 & \ddot{\theta}_0 \end{bmatrix} \begin{bmatrix} \cos(\theta_0 + \lambda_{i0}^*) \\ \sin(\theta_0 + \lambda_{i0}^*) \end{bmatrix} \tag{3.35}$$

and, from Eq. (3.22), the desired relative speed $\dot{q}_{ij}^*$ in Eqs. (3.34) is expressed as

$$\dot{q}_{ij}^* = \rho_{ij}^* \dot{\theta}_0 \begin{bmatrix} -\sin(\theta_0 + \lambda_{ij}^*) \\ \cos(\theta_0 + \lambda_{ij}^*) \end{bmatrix}. \tag{3.36}$$

From Assumption 3.1, exogenous disturbance $d(\theta_0, \dot{\theta}_0, \ddot{\theta}_0, \ddot{q}^*)$ in Eqs. (3.33) is bounded. Convergence property of trajectory X to a ball centered at the origin is determined by analyzing the trajectories of system

$$\dot{\mathrm{X}} = f(\mathrm{X}) + d(\theta_0, \dot{\theta}_0, \ddot{\theta}_0, \ddot{q}^*).$$

To do so, after noticing that

$$\frac{\partial \Delta_2}{\partial \mathrm{X}} \equiv 0,$$

the first variation of Eqs. (3.33) is computed as

$$\delta\dot{\mathrm{X}} = A_{fw}\delta\mathrm{X} + B_{fw}\left(-(H+\Gamma)\,\delta\widetilde{Z} - \frac{\partial \Delta_1(\theta_1, ..., \theta_n)}{\partial \mathrm{X}}\right), \tag{3.37}$$

where

$$\frac{\partial \Delta_1(\theta_1, ..., \theta_n)}{\partial \mathrm{X}} = diag\left(\frac{\partial \Delta_{1,1}(\theta_1)}{\partial \widetilde{\mathcal{Q}}_1}, ..., \frac{\partial \Delta_{n,1}(\theta_n)}{\partial \widetilde{\mathcal{Q}}_n}\right).$$

Since

$$\theta_i = \tan^{-1}\left(\frac{y_i}{x_i}\right)$$

and $q_i = \widetilde{q}_i + q_i^*$, partial derivative

$$\frac{\partial \Delta_{i,1}(\theta_i)}{\partial \widetilde{Q}_i}$$

can be expressed as

$$\frac{\partial \Delta_{i,1}(\theta_i)}{\partial \widetilde{Q}_i} = \left(H_i + \frac{I_2}{n_i}\right) C_i N_i(\dot{q}_i, \theta_i) + \Gamma_i C_i N_{i0}(\dot{q}_i, \theta_i) \tag{3.38}$$

with

$$N_i(\dot{q}_i, \theta_i) = \left[\begin{array}{cc|c} -\frac{\dot{y}_i}{\left\|\dot{q}_i\right\|^2} \sum_{j\in\mathcal{N}_i} \rho_{ij}^* \sin\left(\theta_i + \lambda_{ij}^*\right) & \frac{\dot{x}_i}{\left\|\dot{q}_i\right\|^2} \sum_{j\in\mathcal{N}_i} \rho_{ij}^* \sin\left(\theta_i + \lambda_{ij}^*\right) & \\ \frac{\dot{y}_i}{\left\|\dot{q}_i\right\|^2} \sum_{j\in\mathcal{N}_i} \rho_{ij}^* \cos\left(\theta_i + \lambda_{ij}^*\right) & -\frac{\dot{x}_i}{\left\|\dot{q}_i\right\|^2} \sum_{j\in\mathcal{N}_i} \rho_{ij}^* \cos\left(\theta_i + \lambda_{ij}^*\right) & 0_2 \\ \hline \multicolumn{2}{c|}{0_2} & 0_2 \end{array}\right],$$

$$N_{i0}(\dot{x}_i, \dot{y}_i, \theta_i) = \left[\begin{array}{cc|c} -\frac{\dot{y}_i}{\left\|\dot{q}_i\right\|^2} \rho_{i0}^* \sin\left(\theta_i + \lambda_{i0}^*\right) & \frac{\dot{x}_i}{\left\|\dot{q}_i\right\|^2} \rho_{i0}^* \sin\left(\theta_i + \lambda_{i0}^*\right) & \\ \frac{\dot{y}_i}{\left\|\dot{q}_i\right\|^2} \rho_{i0}^* \cos\left(\theta_i + \lambda_{i0}^*\right) & -\frac{\dot{x}_i}{\left\|\dot{q}_i\right\|^2} \rho_{i0}^* \cos\left(\theta_i + \lambda_{i0}^*\right) & 0_2 \\ \hline \multicolumn{2}{c|}{0_2} & 0_2 \end{array}\right]. \tag{3.39}$$

From Assumption 3.2, bounds (3.21) yield

$$\left|\dot{x}_i\right| \cdot \left\|\dot{q}_i\right\|^{-2} < \frac{v_M}{v_m}, \quad \left|\dot{y}_i\right| \cdot \left\|\dot{q}_i\right\|^{-2} < \frac{v_M}{v_m}, \tag{3.40}$$

which in turn make $N_i(\dot{q}_i, \theta_i)$ and $N_{i0}(\dot{q}_i, \theta_i)$ bounded matrices. It is therefore possible to express $N_i(\dot{q}_i, \theta_i)$ and $N_{i0}(\dot{q}_i, \theta_i)$, for all $i \in \mathcal{V}$, as convex combinations of constant matrices. Let

$$\mathfrak{N} = [N_1, ..., N_n, N_{10}, ..., N_{n0}]$$

and omit arguments in $\dot{q}_i$ and θ_i for brevity. $\mathfrak{N}$ thus lies within the polytope

$$P_{\mathfrak{N}} = Co\left\{\mathfrak{N}_1^1, ..., \mathfrak{N}_1^{n_1}, ..., \mathfrak{N}_i^1, ..., \mathfrak{N}_i^{n_i}, ..., \mathfrak{N}_n^1, ..., \mathfrak{N}_n^{n_n}\right\};$$

that is, $\mathfrak{N} \in P_{\mathfrak{N}}$ means there exist $\alpha_i^k > 0$ such that

$$\begin{array}{c} \mathfrak{N} = \sum_{i=1}^n \sum_{k=1}^{n_i} \alpha_i^k \mathfrak{N}_i^k, \\ \sum_{i=1}^n \sum_{k=1}^{n_i} \alpha_i^k = 1. \end{array} \tag{3.41}$$

The preceding development is required to obtain conditions such that contraction occurs at each vertex of the polytope. Such conditions refer to the uniform negative definiteness of the generalized Jacobian of the error dynamics in Eq. (3.33).

Assume $d \equiv 0$. This occurs if the leader perfectly tracks a straight line, for example. If one succeeds to devise conditions under which $\widetilde{\mathcal{D}}$ is a contraction region, then any trajectory X that starts in $\widetilde{\mathcal{D}}$ converges exponentially to the origin. If the leader, however, tracks a trajectory with some error, i.e. d is not identically zero, X converges exponentially, under the same contraction conditions, to a ball B_d centered at the origin provided B_d is inside $\widetilde{\mathcal{D}}$. The latter remains true if the trajectory of the leader is such that upper bounds on $\left|\dot{\theta}_0\right|$, $\left|\ddot{\theta}_0\right|$, $\left\|\ddot{q}^*\right\|$, and $\left\|\widetilde{Z}_0\right\|$ are sufficiently small. This means that the actual trajectory of the leader cannot vary abruptly.

To better understand Theorem 3.1 which follows, let us define the following:

$$\Lambda_{ij} = \begin{bmatrix} \frac{K_j}{o_j} & 0_2 \\ -H_i - \frac{I_2}{n_i^*} & \frac{H_i+\gamma_i K_i}{2} \end{bmatrix}, \ \Upsilon_{ij} = \begin{bmatrix} \frac{(n_j+n_j^*)H_j+\gamma_j K_j}{2o_j} & 0_2 \\ -H_i - \frac{I_2}{n_i} & \frac{H_i+\gamma_i K_i}{2} \end{bmatrix}. \quad (3.42)$$

Theorem 3.1. ***Convergence of State Trajectories*** [115]. *Suppose the leader unicycle model is in closed loop with controllers defined by (3.25), with $i = 0$, and (3.28), where C_0 is selected such that A_0 in (3.32) is Hurwitz. Assume each follower node i in $\mathcal{V}$ is in closed loop with control laws (3.25) and (3.26), as described in Definition 3.2. All C_i, $i \in \mathcal{V}$, satisfy for some matrix $P = P^T > 0$ and $\lambda_P > 0$ the following matrix inequality*

$$\begin{bmatrix} P\mathcal{A}\left(\mathfrak{N}_i^k\right) + \mathcal{A}\left(\mathfrak{N}_i^k\right)^T P & PB_{fw} \\ B_{fw}^T P & 0 \end{bmatrix} < \begin{bmatrix} -\lambda_P P & C_{fw} \\ C_{fw}^T & (H+\Gamma)^{-1} + (H+\Gamma)^{-T} \end{bmatrix}, \quad (3.43)$$

at each vertex $\mathfrak{N}_i^k$ of $P_{\mathfrak{N}}$, and for all $\mathrm{X} \in \widetilde{\mathcal{D}}$. Furthermore,

$$\mathcal{A}\left(\mathfrak{N}_i^k\right) = A_{fw} - B_{fw}\frac{\partial \Delta_1(\theta_1, ..., \theta_n)}{\partial \mathrm{X}} \quad (3.44)$$

is calculated in Eqs. (3.38) and (3.39), and depends linearly on $\mathfrak{N}_i^k$, provided bounds (3.40) are satisfied. Finally, consider SPR matrices H_i and K_i in control law (3.26) that satisfy the feasible set of matrix inequalities

$$\begin{aligned} \Lambda_{ij} + \Lambda_{ij}^T > 0, \ i \in \mathcal{V}, j \in \mathcal{N}_i^*, \\ \Upsilon_{ij} + \Upsilon_{ij}^T > 0, \ i \in \mathcal{V}, j \in N_i, \end{aligned} \quad (3.45)$$

where $\mathcal{N}_i^*$ *and* N_i *are respectively defined in Eqs. (3.18) and (3.19), or more conservatively the set of inequalities*

$$\begin{aligned} &k_{1j} > o_j \frac{\left(h_{1i}+\frac{1}{n_i^*}\right)^2 + h_{2i}^2}{2(h_{1i}+\gamma_i k_{1i})},\ j \in \mathcal{N}_i^*, \\ &\left(n_j + n_j^*\right) h_{1j} + \gamma_j k_{1j} > o_j \frac{\left(h_{1i}+\frac{1}{n_i}\right)^2 + h_{2i}^2}{(h_{1i}+\gamma_i k_{1i})},\ j \in N_i, \end{aligned} \quad (3.46)$$

when H_i *and* K_i *are selected as diagonal matrices* $h_i I_2$ *and* $k_i I_2$*. Operators* h_i *and* k_i *are linear causal SPR, expressed in complex notation as* $h_{1i}+jh_{2i}$*, and* $k_{1i} + jk_{2i}$*, respectively, and map signals from* L_p *to* L_p*,* $p \in \{2, \infty\}$*.*

Then, for sufficiently small upper bounds of $\left|\dot{\theta}_0\right|$, $\left|\ddot{\theta}_0\right|$, $\left\|\ddot{q}^*\right\|$ *and* $\left\|\widetilde{Z}_0\right\|$*, state* X *of Eq. (3.33), which expresses the tracking error in position and speed of each vehicle, converges exponentially to a ball* B_d *centered at zero and with radius* R_{B_d} *defined by*

$$R_{B_d} = \frac{d^M}{|\lambda_P|}, \quad (3.47)$$

where

$$d^M = \sup_{\theta_0, \dot{\theta}_0, \ddot{\theta}_0, \ddot{q}^*} \left\| \Omega d(\theta_0, \dot{\theta}_0, \ddot{\theta}_0, \ddot{q}^*) \right\| .$$

and $P = \Omega^T \Omega$.

We introduce the following notation before considering Lemma 3.1 used in the proof of Theorem 3.1. For a matrix given as

$$M = \begin{bmatrix} A & B \\ C & D \end{bmatrix},$$

where A, B, C and D are 2×2 matrices, let

$$[M]_{ilkj} = \begin{bmatrix} [A]_{ik} & [B]_{ij} \\ [C]_{lk} & [D]_{lj} \end{bmatrix} = \begin{array}{c} \begin{matrix} \ \ \ k\text{th} & j\text{th} \ \ \ \end{matrix} \\ \begin{bmatrix} \ddots & \vdots & & \vdots & \\ \dots & A & \dots & B & \dots \\ & \vdots & \ddots & \vdots & \\ \dots & C & \dots & D & \dots \\ & \vdots & & \vdots & \ddots \end{bmatrix} \end{array} \begin{matrix} \\ i\text{th} \\ \\ l\text{th} \\ \\ \end{matrix} \quad (3.48)$$

represent any $2n \times 2n$ block matrix, where $n \geq 2$, with entries being zero except for the blocks A, B, C and D shown in Eq. (3.48).

Lemma 3.1. ***SPR*** $H+\Gamma$ [115]. *If the SPR matrices* H_i *and* K_i *in control law (3.26) satisfy the feasible set of LMIs (3.45) or more conservatively (3.46) in case of diagonal* H_i *and* K_i*, then* $H+\Gamma$*, which defines the map* $\widetilde{Z} \mapsto I_{H+\Gamma}$*, is strictly positive real.*

Proof. From Definition 3.1, since $\mathcal{G}$ and $\mathcal{G}_0$ are DAG, the set of vertices $\mathcal{V}$ can be topologically sorted. Establishing a bijection from the sorted set V to the set $\{\widetilde{Z}_i \in \mathbb{R}^2 \mid i \in V\}$ such that $i \to \widetilde{Z}_i$ results in a topologically sorted $\widetilde{Z}$, which is shown in Fig. 3.14. From now on, it is assumed that $\mathcal{V}$ and $\widetilde{Z}$ are sorted. Therefore, $H+\Gamma$ can be expressed as a lower triangular matrix parameterized by $H_i, \frac{1}{n_i}, o_i$ and K_i. From Definition 3.2 and Eqs. (3.31), $H+\Gamma$ can be described as

$$H+\Gamma = \sum_{i=2}^{n} \left(\sum_{j\in\mathcal{N}_i^*} \begin{bmatrix} \left[\frac{K_j}{o_j}\right]_{jj} & [0_2]_{ji} \\ \left[-H_i - \frac{I_2}{n_i^*}\right]_{ij} & [H_i+\gamma_i K_i]_{ii} \end{bmatrix} + \sum_{j\in N_i} \begin{bmatrix} [0_2]_{jj} & [0_2]_{ji} \\ \left[-H_i - \frac{I_2}{n_i}\right]_{ij} & [H_i+\gamma_i K_i]_{ii} \end{bmatrix} \right). \tag{3.49}$$

On the right-hand side of Eq. (3.49), the matrix formed by the first sum in j expresses the fact that each neighbor j of i is connected to the leader, while node i is possibly but not necessarily connected to the leader. The matrix formed by the second sum in j simply expresses the connection $j \rightsquigarrow i$ and possibly $0 \rightsquigarrow i$. Recall that connections to the leader, as specified in Definition 3.1, must at least ensure the weak connectedness of $\mathcal{G}_0$, which is equivalent to the following condition:

$$rank\,(H+\Gamma) = 2n.$$

From the connectedness assumption and the topological sorting of the nodes, it is possible to rewrite $H+\Gamma$ as

$$H+\Gamma = \sum_{i=2}^{n} \left(\sum_{j\in\mathcal{N}_i^*} [\Lambda_{ij}]_{jiji} + \sum_{j\in N_i} [\Upsilon_{ij}]_{jiji} \right) + \sum_{i\in\mathcal{O}} [\Xi_{ij}]_{jiji}\,, \tag{3.50}$$

where $[\Lambda_{ij}]_{jiji}$ and $[\Upsilon_{ij}]_{jiji}$ are obtained from Eqs. (3.42) using the matrix notation introduced in Eq. (3.48). Furthermore,

$$[\Xi_{ij}]_{jiji} = \begin{bmatrix} [0_2]_{jj} & [0_2]_{ji} \\ [0_2]_{ij} & \left[\frac{(n_i+n_i^*)H_i+\gamma_i K_i}{2}\right]_{ii} \end{bmatrix},$$

where j is any integer such that $j \le n$ and $j \ne i$.

Operator $H + \Gamma$ is SPR if

$$Q = \widetilde{Z}^T(H + \Gamma + \widehat{H}^T + \widehat{\Gamma}^T)\widetilde{Z} > 0$$

for all $\widetilde{Z} \neq 0$. When the symmetric part of Λ_{ij} and Υ_{ij}, expressed by LMIs (3.45), is SPR, the last inequality is satisfied. From the Schur complement lemma, LMIs (3.45) are converted into

$$\begin{aligned} &\tfrac{K_j}{o_j} - \tfrac{1}{4}(H_i + \tfrac{I_2}{n_i^*})(\tfrac{H_i+\gamma_i K_i}{2})^{-1}(H_i + \tfrac{I_2}{n_i^*}) > 0,\ i \in \mathcal{V}, j \in \mathcal{N}_i^* \\ &\tfrac{(n_j+n_j^*)H_j+\gamma_j K_j}{2o_j} - \tfrac{1}{4}(H_i + \tfrac{I_2}{n_i^*})(\tfrac{H_i+\gamma_i K_i}{2})^{-1}(H_i + \tfrac{I_2}{n_i^*}) > 0,\ i \in \mathcal{V}, j \in N_i. \end{aligned} \tag{3.51}$$

One can thus conclude that feasible solutions of LMIs (3.45) exist, provided H_i and K_i are selected as diagonal matrices $h_i I_2$ and $k_i I_2$, respectively. Such matrices applied to (3.51) lead to conditions expressed as inequalities (3.46) for which solutions h_j and k_j always exist. In fact, h_j and k_j can be computed in descending order; that is, from the terminal nodes to the nodes that are neighbors of the leader. First, positive h_i and k_i are selected for nodes $i \in \mathcal{O}$. These so-called slack parameters can be assigned any value as long as they are positive. The positivity of h_i and k_i, for nodes $i \in \mathcal{O}$, ensures convergence of the trajectories. Then, h_j and k_j $(j \rightsquigarrow i)$ are computed iteratively from inequalities (3.46). This procedure is carried out until all the h_j and k_j have been calculated [114].

Conditions (3.45) or more conservatively (3.46), and the SPR property of H_i and K_i express the positive definiteness of the symmetric part of matrices Λ_{ij} and Υ_{ij} in Eq. (3.50), and of the (i,i) entry of Ξ_{ij}. Each quadratic term of Q is bounded from above and below by $\lambda_m \left\|[\widetilde{Z}_j^T\ \widetilde{Z}_i^T]\right\|^2$ and $\lambda_M \left\|[\widetilde{Z}_j^T\ \widetilde{Z}_i^T]\right\|^2$, respectively. The smallest eigenvalue is λ_m, and is greater than zero, while the largest eigenvalue is λ_M for matrix

$$\Lambda_{ij} + \Lambda_{ij}^T, \Upsilon_{ij} + \Upsilon_{ij}^T$$

or

$$\frac{(n_i + n_i^*)\, H_i + \gamma_i K_i}{2}$$

when $i \in \mathcal{O}$. In the latter case, quadratic bounds simplify to $\lambda_m ||\widetilde{Z}_i||^2$ and $\lambda_M ||\widetilde{Z}_i||^2$. Hence, as the sum in Eq. (3.50) spans the whole graph $\mathcal{G}$, it is possible to find $\delta_m > 0$ and $\delta_M > 0$ such that

$$\delta_m ||\widetilde{Z}||^2 \leq Q \leq \delta_M ||\widetilde{Z}||^2,$$

which in turn guarantees that $H + \Gamma$ is SPR. □

Proof. The proof of Theorem 3.1 is given as follows. Time derivative of the squared length $\delta X^T P \delta X$ along trajectory of Eq. (3.37) leads to

$$\frac{d}{dt}\left(\delta X^T P \delta X\right) = \delta X^T Q \delta X - 2\delta X^T P B_{fw} \delta I_{H+\Gamma},$$
$$Q = -\left(P\left(A_{fw} - B_{fw}\frac{\partial\Delta_1(\theta_1,\ldots,\theta_n)}{\partial X}\right) + \left(A_{fw} - B_{fw}\frac{\partial\Delta_1(\theta_1,\ldots,\theta_n)}{\partial X}\right)^T P\right). \tag{3.52}$$

Since X lies in $\widetilde{\mathcal{D}}$, matrix Q linearly depends on $\mathfrak{N}_i^k$ and thus varies within a polytope that is closely related to $P_{\mathfrak{N}}$; that is, there exist $\alpha_i^k > 0$ and $Q_i^k\left(\mathfrak{N}_i^k\right)$ such that

$$Q = \sum_{i=1}^n \sum_{k=1}^{n_i} \alpha_i^k Q_i^k\left(\mathfrak{N}_i^k\right),$$
$$\sum_{i=1}^n \sum_{k=1}^{n_i} \alpha_i^k = 1, \tag{3.53}$$

which yields

$$\frac{d}{dt}\left(\delta X^T P \delta X\right) = -\delta X^T \sum_{i=1}^n \sum_{k=1}^{n_i} \alpha_i^k Q_i^k\left(\mathfrak{N}_i^k\right)\delta X - 2\delta X^T P B_{fw} \delta I_{H+\Gamma}$$
$$= \sum_{i=1}^n \sum_{k=1}^{n_i} \alpha_i^k \begin{bmatrix} \delta X \\ -\delta I_{H+\Gamma} \end{bmatrix}^T \begin{bmatrix} -Q_i^k\left(\mathfrak{N}_i^k\right) & P B_{fw} \\ B_{fw}^T P & 0 \end{bmatrix} \begin{bmatrix} \delta X \\ -\delta I_{H+\Gamma} \end{bmatrix}. \tag{3.54}$$

Note that a necessary condition for inequality (3.43) to hold is that

$$(H+\Gamma)^{-1} + (H+\Gamma)^{-T} > 0,$$

which follows from Lemma 3.1. Hence, from inequality (3.43), time derivative given by Eq. (3.54) becomes

$$\frac{d}{dt}\left(\delta X^T P \delta X\right) \leq \begin{bmatrix} \delta X \\ -\delta I_{H+\Gamma} \end{bmatrix}^T \begin{bmatrix} -\lambda_P & C_{fw} \\ C_{fw}^T & (H+\Gamma)^{-1} + (H+\Gamma)^{-T} \end{bmatrix} \begin{bmatrix} \delta X \\ -\delta I_{H+\Gamma} \end{bmatrix}. \tag{3.55}$$

By noticing

$$\delta I_{H+\Gamma}^T\left(C_{fw} + C_{fw}^T\right)\delta X = \delta I_{H+\Gamma}^T\left((H+\Gamma)^{-1} + (H+\Gamma)^{-T}\right)\delta I_{H+\Gamma},$$

inequality (3.55) simplifies to

$$\frac{d}{dt}\left(\delta X^T P \delta X\right) \leq -\lambda_P \left\|\delta X\right\|^2, \tag{3.56}$$

which proves that $\widetilde{\mathcal{D}}$ is a contraction region. Hence, trajectories of the unexcited system (3.33) converge to zero. By unexcited system, we mean signal d is equal to zero in Eq. (3.33). The conclusion remains the same if the tracking error of the leader, $\widetilde{Z}_0$, asymptotically approaches zero.

Consider the case where d is not identically zero. From Ref. [106], p. 688, $X \in \widetilde{\mathcal{D}}$ converges exponentially to a ball B_d centered at 0 and of radius $d^M/|\lambda_P|$, which remains entirely in the contraction region $\widetilde{\mathcal{D}}$ if $\left|\dot{\theta}_0\right|, \left|\ddot{\theta}_0\right|, \left\|\ddot{q}^*\right\|$ and $\left\|\widetilde{Z}_0\right\|$ are sufficiently small since $\left\|d(\theta_0, \dot{\theta}_0, \ddot{\theta}_0, \ddot{q}^*)\right\|$ can be bounded affinely in $\left|\dot{\theta}_0\right|, \left|\ddot{\theta}_0\right|, \left\|\ddot{q}^*\right\|$ and $\left\|\widetilde{Z}_0\right\|$ as follows

$$\left\|d(\theta_0, \dot{\theta}_0, \ddot{\theta}_0, \ddot{q}^*)\right\| \leq \|\Gamma\| \left\|\widetilde{Z}_0\right\| + \left\|\ddot{q}^*\right\|$$
$$+ \sup_{i\in\mathcal{V}} \left(\rho_{i0}^* \left(\left|\dot{\theta}_0^2\right| + \left|\ddot{\theta}_0\right|\right) + \sum_{j\in\mathcal{N}_i} \left\|\left(H_i + \tfrac{I_2}{n_i}\right) C_i\right\| \left|\rho_{ij}^*\dot{\theta}_0\right| + \|\Gamma_i C_i\| \left|\rho_{i0}^*\dot{\theta}_0\right|\right). \tag{3.57}$$

□

A few remarks on Theorem 3.1 are in order. First, analytical feasibility of BMIs (3.43) is not readily provable, especially if the number of vertices is large. However, the number of vertices at which BMIs must be verified can be reduced if K_i is equal to $H_i + \frac{I_2}{n_i}$, which is always possible from inequalities (3.46). In this case, Eq. (3.38) becomes

$$\partial\Delta_{i,1}(\theta_i)/\partial\widetilde{Q}_i = \Gamma_i C_i \mathrm{N}_i(\dot{q}_i, \theta_i),$$

where

$$\mathrm{N}_i(\dot{q}_i, \theta_i) = N_i(\dot{q}_i, \theta_i) + N_{i0}(\dot{q}_i, \theta_i).$$

Matrix N_i is characterized by four nonzero elements that can be bounded from above. Matrix N_i thus varies within a polytope of eight vertices. Second, from Eqs. (3.39), bounds on the nonzero entries of N_i depend on a term of the form $\rho_{ij}^*\dot{y}_i \left\|\dot{q}_i\right\|^{-2}$, which is bounded from above by $\rho_{ij}^*/\left\|\dot{q}_i\right\|$. Hence, commands on the geometry of the formation, where at some time instant all the ρ_{ij}^* are multiplied by a large factor, may require increasing the speed $\left\|\dot{q}_i\right\|$ of vehicle i so that $\mathfrak{N}$ remains in polytope $P_{\mathfrak{N}}$. Augmenting the speed of the leader results in an increase in $\left\|\dot{q}_i\right\|$. Third, speed tracking error transients resulting from a step in ρ_{ij}^* must remain within $\widetilde{\mathcal{D}}$ otherwise $\mathfrak{N}$ goes outside polytope $P_{\mathfrak{N}}$, which violates the contraction conditions. Therefore, to achieve changes in the geometry of the formation by means of step commands in relative positions, given by ρ_{ij}^*, requires low-pass filtering of such commands so that the state remains within $\widetilde{\mathcal{D}}$, as illustrated in the simulations of Sec. 3.6.1.

3.2.6 *Modeling a formation of UAVs with realistic nonlinear dynamics*

Even though the unicycle model can represent the motion of a variety of vehicles equipped with autopilots and servomechanisms, it makes sense to close the loop with a more detailed mathematical model representative of actual vehicle dynamics. This section shows how to obtain a tractable mathematical model which is later used for the design of decentralized fault detectors. First, assume that every vehicle in the formation is equipped with fine-tuned autopilots so that the first-order unicycle-type approximation (3.20) remains valid for a prescribed set of operating conditions. Second, rewrite the outer-loop control of a follower vehicle i with state (x_i, y_i, z_i), given in Definition 3.2, as

$$\begin{bmatrix} u_{xi} \\ u_{yi} \end{bmatrix} = \underbrace{-h_i \sum_{j\in N_i} k_i \begin{bmatrix} x_i \\ y_i \end{bmatrix}}_{=[u_{1,xi}, u_{1,yi}]^T} + \underbrace{h_i \sum_{j\in N_i} k_i \begin{bmatrix} x_j - x_{ij}^* \\ y_j - y_{ij}^* \end{bmatrix}}_{=[u_{2,xi}, u_{2,yi}]^T}, \qquad (3.58)$$

$$\begin{bmatrix} x_{ij}^* \\ y_{ij}^* \end{bmatrix} = \rho_{ij}^* \begin{bmatrix} \cos(\theta_i + \lambda_{ij}^*) \\ \sin(\theta_i + \lambda_{ij}^*) \end{bmatrix}.$$

In Eqs. (3.58), h_i is a linear, causal operator. In practice, for the ALTAV and quadrotor systems, h_i can be selected as a first-order biproper transfer function. Gain k_i can be set to unity for all practical purposes. Autopilot commands u_{xi} and u_{yi} control the $x-y$ plane motion of the vehicle. As an example, the reader is referred to the ALTAV autopilot equations in (3.2). Set-point regulation in z is obtained by PID control around a prescribed elevation z_t, which is assumed to be known by the vehicles prior to mission.

Third, the dynamics of a vehicle in closed loop with its autopilots and with outer-loop control $[u_{1,xi}, u_{1,yi}]^T$ is approximated by a parameter-dependent linear system

$$\dot{q}_i = A_i(\alpha_i)q_i + B_i(\alpha_i) \begin{bmatrix} u_{2,xi} \\ u_{2,yi} \\ z_t \end{bmatrix}, \qquad (3.59)$$

$$\begin{bmatrix} x_i \\ y_i \\ z_i \end{bmatrix} = \begin{bmatrix} 1\,0\,0\,0\,0\,0 \\ 0\,0\,1\,0\,0\,0 \\ 0\,0\,0\,0\,0\,1 \end{bmatrix} q_i,$$

where x_i, y_i, and z_i represent vehicle translations, and state variable q_i is in $\mathbb{R}^{n\times 1}$. The order n of the system is determined such that the parameter identification of Eqs. (3.59) provides a satisfactory approximation of the closed-loop dynamics of the vehicle. The parameter vector α_i depends on controller gains h_i and k_i. Control signals $u_{2,xi}$ and $u_{2,yi}$ represent the couplings with neighbors of vehicle i. Signals $u_{2,xi}$, $u_{2,yi}$, and z_t excite the system for the purpose of parameter identification, in which case signals $u_{2,xi}$ and $u_{2,yi}$ are assigned to typical trajectories such as square and sinusoidal waves, whereas elevation z_t is kept constant. It is important to mention that control gains usually have to be tuned; for example, by means of numerical simulations. Commercially available software such as MATLAB® System Identification Toolbox™ [119] can be used to calculate matrices $A_i(\alpha_i)$ and $B_i(\alpha_i)$. The model given by Eqs. (3.59) is general enough to represent a variety of unmanned aerial vehicles equipped with control laws $[u_{1,xi}, u_{1,yi}]^T$, including 6-DOF nonlinear dynamic models of ALTAV and quadrotor, presented in Sec. 3.1 and used in Sec. 3.6 to test the proposed health monitoring and adaptation algorithms.

3.2.7 *Trajectory of the leader and obstacle/threat avoidance*

Leader trajectories may be pre-computed for a specific mission plan, and be re-generated online to accommodate dynamic events such as the presence of obstacles and the occurrence of faults/failures, to comply with constraints, and to adapt to changes in the mission. For formations flying at a constant altitude, a trajectory is obtained for the leader in the form of variables x^* and y^* on the $x - y$ plane, as stated in Assumption 3.1. A pair (x^*, y^*) is defined for all or a section of the mission space, and over some time interval. The trajectory must be flyable; namely, the trajectory represented by the pair should satisfy inherent kinematic and dynamic limits of the vehicle. Furthermore, to ensure safe and reliable missions, UAVs should be capable of performing relatively quick adaptation to dynamic events in addition to being responsive to commands issued by the operating crew. Automatic trajectory re-generation is illustrated by means of an example, for brevity. Consider Fig. 3.15, which represents the top view of a formation in level flight and within a constrained area. A leader (L) is required to reach waypoints 1 and 2, labeled as WP1 and WP2, respectively, with poses identified by triangles. Specific poses may be required for instance when a vehicle must capture data on the area and, given its payload characteristics, has to be oriented in specific ways. Rectangles indicate constraints on the allowable flight path. There is a circular obstacle or threat (O) on the pathway. The dashed line T0 is the trajectory originally planned for the leader. However, at the time trajectory T0 was calculated, there was no obstacle. During flight, as soon as timely sensor information on the obstacle is available to the leader, a calculation is made to obtain a second trajectory, labeled as T1, enabling the vehicle to fly around the static obstacle. T1 is obtained by specifying an intermediate waypoint, identified as I in the figure, and located a certain distance perpendicular to the crossing of the obstacle boundary by T0. Obviously, the alternative trajectory must be generated quickly enough so that the vehicle has time to modify its course.

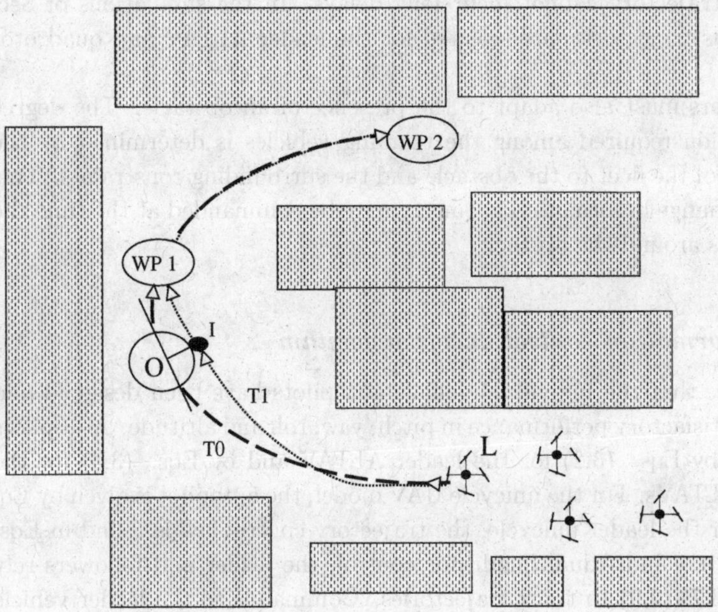

Fig. 3.15 Example of trajectory re-generation for the purpose of collision/threat avoidance.

To obtain feasible and effective trajectories in a relatively short time, one may use so-called Dubins paths, which are collections of lines and arcs resulting in the shortest achievable paths. Dubins paths are relatively simple to compute, and are well suited for fixed-wing UAVs, as the latter require only constant-rate turns to move along such paths. Yet, discontinuous demands during transitions between lines and arcs occur and hence require special attention [120]. An alternative is the use of Pythagorean hodograph (PH) curves [121, 122]. PH trajectories are used in the context of UAV flight in Refs. [113, 123]. The use of PH curves eliminates the issue of discontinuity associated with Dubins paths. The generation of PH trajectories necessitates relatively short computing times. However, there is no guarantee that the PH curves provide shortest-path solutions between waypoints [120]. The computation of the trajectories must be done expeditiously for two reasons. Prior to mission, trajectory generation requires acute, up-to-date knowledge of the theater. Delays in the computing of trajectories may render their use irrelevant. During mission, to allow the UAVs to perform collision/threat avoidance maneuvers, computation of a new trajectory cannot incur long delays. In the simulations of Sec. 3.6, Dubins trajectories are selected for the leader ALTAV and quadrotor vehicles.

Followers must also adapt to the presence of an obstacle. The degree of separation required among the teaming vehicles is determined by the proximity of the fleet to the obstacle and the surrounding constraints. This means a change in formation geometry may be commanded at the time the leader goes around the obstacle.

3.2.8 *Formation control design algorithm*

We assume that the individual vehicle autopilots have been designed and provide satisfactory performance in pitch, yaw, roll and altitude. Autopilots are given by Eqs. (3.2) for the leader ALTAV and by Eqs. (3.4) for the follower ALTAVs. For the unicycle UAV model, the autopilot is given by Eq. (3.25). For the leader unicycle, the trajectory control law is found in Eqs. (3.4). For the individual quadrotor control, the leader and followers rely on Eqs. (3.7)-(3.9) to track trajectories. Commands to the leader vehicle take the form of (x_d, y_d, z_d, t), as supplied by the trajectory generator. For the followers, the formation control law is given in Eqs. (3.58) applied as commands to the low-level control schemes (3.7)-(3.9). We now summarize the design steps of the formation control system.

Step 1. Determine the topology of the information flow in the formation, as detailed in Definition 3.1.

Step 2a. For the unicycle-type follower model and for the given formation topology, obtain controllers C_i, H_i and K_i in (3.26) such that inequalities (3.43) and (3.45), or (3.46), are satisfied. Such selection of controllers ensures convergence of trajectories, according to Theorem 3.1. COTS software LMI tools, such as Scilab [124] and LMI toolbox [108], may be used to calculate the controllers. Alternatively, a designer may choose to obtain simple PID-type controllers, tune them by means of simulations, and then verify that the chosen parameters satisfy the inequalities.

Step 2b. For ALTAV and quadrotor followers, compute controllers h_i and k_i in Eqs. (3.58) building upon values obtained for a similar formation of unicycle. Tuning of the controllers may be required.

Step 3. Include a low-pass filter between online commands on formation geometry issued by the operating crew and actual input commands to the formation control system.

Step 4. Tune the controller gains and filter parameters, if need be, by means of numerical simulations. The designer should excite the formation with various disturbances and commanded geometries.

3.3 Observer-Based Decentralized Abrupt Fault Detector

3.3.1 *Context*

A method for designing control laws for leader-to-follower formations is available for nominal operating conditions and is described in Sec. 3.2. In practice, such formation control scheme does not guarantee safe and reliable multi-vehicle operation when a fault or failure occurs on a flight-critical component of any of the vehicles, or when a vehicle is damaged, while communications are lost intermittently. Limited recovery provided by the FTC system of a faulty vehicle may adversely affect the motion of the other vehicles in the formation, causing unsafe flight and possibly mission failure. To handle such off-nominal conditions, an additional system, known as the cooperative health management system (CHM), should be embedded onboard the UAVs.

A few research results are currently available on the topic of reliability and safety in unmanned vehicle formation flight. Here, we cite some of those results. Formation reconfiguration under information flow faults is presented in Refs. [125–127], where graph theory and a modified Dijkstra algorithm are utilized to command geometry changes and to optimally reconfigure communication channels once information flow faults have been detected. From another perspective, Ref. [128] proposes an interacting-multiple-model FDI approach for formations faced with communications failure. A command-and-control architecture for the deployment of multiple aerial vehicles is proposed in Ref. [129]. Centralized mission planning and health management are performed on the ground and in real time. In

Ref. [130], a semi-decentralized fault detector is designed for precise formation flight of satellites. The fault detector exploits variables transmitted from one neighbor to another via a communication network. For formations of ground vehicles, decentralized model-based abrupt fault diagnosis is detailed in Ref. [131]. Fault diagnosis builds upon bond graph modeling and gives particular attention to fault propagation dynamics among robots. Fault detectors are developed in Ref. [132], where measurements of local variables are used in combination with an overlapping decomposition of the initial large-scale system. This approach is applied to the health monitoring of a string of five interconnected tanks. A distributed game-theoretic fault detector is proposed in Ref. [133] with application to a platoon of cars. Finally, a decentralized fault detector for a platoon-type collection of linear, time-invariant subsystems is presented in Ref. [134], where a filter is designed for each subsystem. State estimates are shared among the subsystems so that dynamic couplings can be exploited.

Here, an observer-based decentralized abrupt fault detector, referred to as DAFD, is presented. DAFD addresses the following situation: a formation is faced with concurrent communications loss and component-level faults in one or more follower vehicles. The basic principle of DAFD is explained in Fig. 3.16. The state of the formation cannot be published across the network at the time of faults. Thus, sensor data available onboard each UAV are used as inputs to an observer. The information obtained from the sensors acts as the redundancy needed to accomplish fault detection. In each UAV, a robust observer based on the use of a reduced-order, uncertain, linearly parameterized model of neighboring UAVs outputs a number of residues. When the magnitude of the residues crosses a user-defined threshold for some time time, a faulty behavior is detected. The CHM function then commands the UAV to stop following the faulty predecessor UAV, and instead follow the closest operational UAV within range, if any.

3.3.2 *Simplified model of vehicle closed-loop dynamics*

We propose to develop a simplified state-space representation of neighboring vehicle dynamics in closed loop with their control law and to design DAFD based on this simplified model. It is assumed that a vehicle is not considered faulty by its neighbors as long as it can achieve position trajectory tracking using the measured relative distances regardless of its attitude. The attitude of the 6-DOF model of the vehicle is therefore irrelevant for the design of DAFD. Substituting $u_{xi} - u_{1,xi}$ and $u_{yi} - u_{1,yi}$ in Eqs. (3.58)

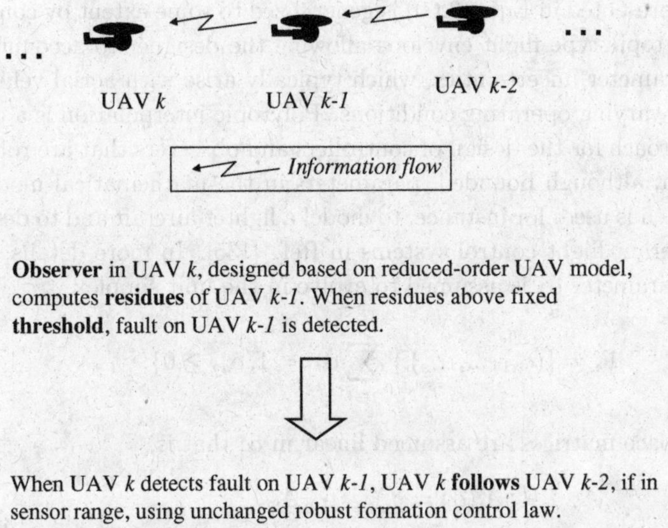
...
...
UAV k
UAV k-1
UAV k-2
Information flow
Observer in UAV k, designed based on reduced-order UAV model, computes **residues** of UAV k-1. When residues above fixed **threshold**, fault on UAV k-1 is detected.
When UAV k detects fault on UAV k-1, UAV k **follows** UAV k-2, if in sensor range, using unchanged robust formation control law.

Fig. 3.16 Basic principle of decentralized abrupt fault detection and adaptation.

for $u_{2,xi}$ and $u_{2,yi}$ in Eqs. (3.59), respectively, yields the simplified model of follower vehicle $i \in \mathcal{V}$ given as

$$\dot{q}_i = A_i(\alpha_i)q_i + B_i(\alpha_i) \underbrace{\begin{bmatrix} h_i \sum_{j \in N_i} k_i(x_j - x_{ij}^*) \\ h_i \sum_{j \in N_i} k_i(y_j - y_{ij}^*) \\ z_t \end{bmatrix}}_{=v_{i1}-v_{i2}} \tag{3.60}$$

$$\begin{bmatrix} x_i \\ y_i \\ z_i \end{bmatrix} = \begin{bmatrix} 1\,0\,0\,0\,0\,0 \\ 0\,0\,1\,0\,0\,0 \\ 0\,0\,0\,0\,0\,1 \end{bmatrix} q_i$$

where

$$\begin{bmatrix} x_{ij}^* \\ y_{ij}^* \end{bmatrix} = \rho_{ij}^* \begin{bmatrix} \cos(\lambda_{ij}^* + \psi_i) \\ \sin(\lambda_{ij}^* + \psi_i) \end{bmatrix};$$

x_i, y_i, and z_i are the vehicle translations; ψ_i is the heading angle of vehicle i; and z_t is the prescribed altitude of the formation, which is assumed to be known prior to mission or be a piecewise constant command published by the leader during mission. State variable q_i is a vector in $\mathbb{R}^{n \times 1}$, with $n = 6$. The entries of the state-space vector q_i are labeled as follows:

$$q_i = [q_{i1}, q_{i2}, q_{i3}, q_{i4}, q_{i5}, q_{i6}]^T.$$

The model presented in Eqs. (3.60) is generalized to some extent by considering a polytopic-type flight envelope allowing the designer to account for possible parameter uncertainties, which typically arise with aerial vehicles due to time-varying operating conditions. Polytopic interpolation is a convenient approach for the design of controllers and observers that are robust to uncertain, although bounded, parameters in the mathematical models. This approach is used, for instance, to model a fighter aircraft and to design its fault-tolerant flight control systems in Ref. [135]. In more details, the uncertain parameter α_i is assumed to evolve in the unit simplex

$$\Gamma_i = \{(\alpha_{i1}, ..., \alpha_{is}) \mid \sum_{j=1}^{s} \alpha_{ij} = 1, \alpha_{ij} \geq 0\}.$$

The state-space matrices are assumed linear in α; that is,

$$\begin{aligned} A_i(\alpha) &= \textstyle\sum_{j=1}^{s} \alpha_{ij} A_{ij}, \\ B_i(\alpha) &= \textstyle\sum_{j=1}^{s} \alpha_{ij} B_{ij}. \end{aligned} \tag{3.61}$$

Note that $A_i(\alpha)$ (resp.$B_i(\alpha)$) can be decomposed as the sum of a nominal matrix $A_i^* = A_i(\alpha^*)$ (resp. $B_i^* = B_i(\alpha^*)$) and a deviation matrix $\widetilde{A}_i(\alpha)$ (resp.$\widetilde{B}_i(\alpha)$). The latter evolves in the same polytope as that of $A_i(\alpha)$ (resp.$B_i(\alpha)$); that is,

$$\begin{aligned} \widetilde{A}_i(\alpha) &= \textstyle\sum_{j=1}^{s} \alpha_{ij} \widetilde{A}_{ij}, \\ \widetilde{B}_i(\alpha) &= \textstyle\sum_{j=1}^{s} \alpha_{ij} \widetilde{B}_{ij}. \end{aligned} \tag{3.62}$$

Equations (3.60) and (3.61) constitute the simplified model used to derive DAFD. The simplified model is depicted in Fig. 3.17(b) for the case of a three-vehicle formation. The more elaborate mathematical model of the UAS and of the formation of three vehicles is shown in Fig. 3.17(a), which includes the low-level control blocks. The design of DAFD does not require such detailed modeling. In fact, the design of DAFD depends on the simplified model of vehicle i. For the purpose of health monitoring, each vehicle is expected to utilize an observer that is tuned from the knowledge of $A_i(\alpha)$ and $B_i(\alpha)$. Yet, a polytopic representation may be used, although with some degree of conservatism, to derive a single DAFD function for the entire homogeneous fleet of UAVs. Every vehicle of the formation is then equipped with the same observer. To carry out such design, a single polytope represents $A_i(\alpha)$ and $B_i(\alpha)$ for all $i \in \mathcal{V}$. This approach is shown in Sec. 3.6 to result in satisfactory performance.

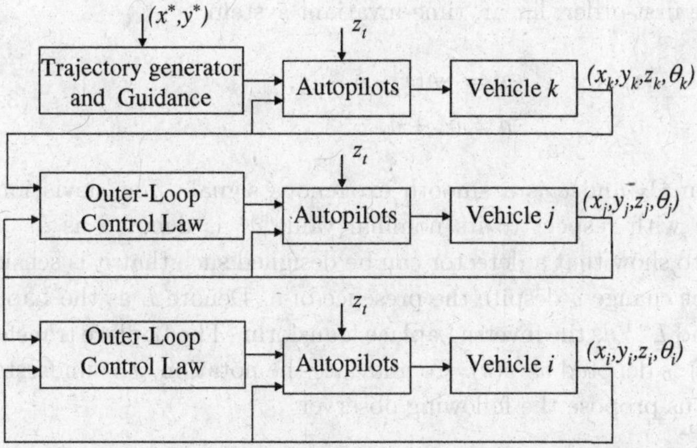

(a)

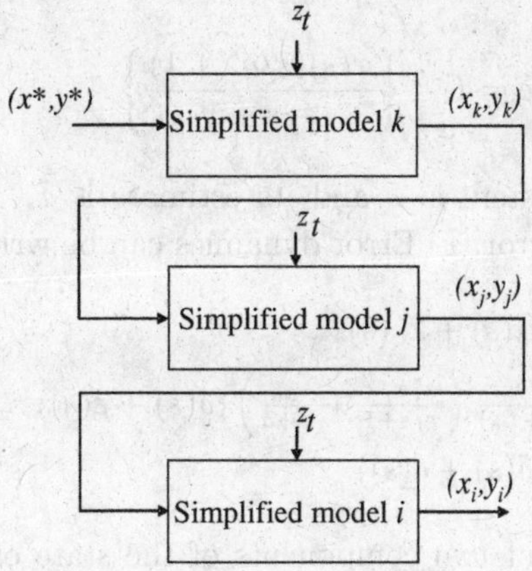

(b)

Fig. 3.17 Block diagrams for (a) detailed and (b) simplified models of a three-UAS formation. Lines indicate information flow among the vehicles, as required by the formation control law.

3.3.3 *Concept of observer with disturbance attenuation*

As a preliminary step in the design of DAFD, we illustrate the underlying idea of an observer whose goal is to provide a residue that is sensitive to faulty signals despite the presence of unknown exogenous disturbances.

Consider the first-order, linear, time-invariant system

$$\begin{aligned} \dot{x} &= -a(x - \delta - e), \\ a &= a^* + \widetilde{a}, \end{aligned} \tag{3.63}$$

where δ is in Ω_δ and e is a smooth exogenous signal. The deviation of parameter a with respect to its nominal value a^* is denoted as $\widetilde{a}$. The objective is to show that a detector can be designed such that it is sensitive to the abrupt change δ despite the presence of e. Denote L as the Laplace transform and L^{-1} as the inverse Laplace transform. The Laplace transform of signal $x(t)$ is denoted as $x(s)$. To alleviate the notation, x is understood as $x(t)$. Let us propose the following observer

$$\dot{\widehat{x}} = -a^*(\widehat{x} - \widehat{\delta} - \widehat{e}) - L_o\widetilde{x}, \tag{3.64}$$

where $\widetilde{x} = \widehat{x} - x$, L_o is the observer gain, and

$$\widehat{\delta} + \widehat{e} = L^{-1}\left\{\frac{x(s)(s/a^* + 1)}{\tau_f s + 1}\right\}. \tag{3.65}$$

The true state of the system is x and its estimate is $\widehat{x}$. The difference between $\widehat{x}$ and x is the error, $\widetilde{x}$. Error dynamics can be written as

$$\begin{aligned} \widetilde{x}(s) &= \widetilde{x}_\delta(s) + \widetilde{x}_e(s) \\ &= \tfrac{-1}{(s+a^*+L_o)}\left(\tfrac{a^*\tau_f s}{\tau_f s+1} + \tfrac{\widetilde{a}s}{s+a}\right)(\delta(s) + e(s)) \\ &= H(s)(\delta(s) + e(s)), \end{aligned} \tag{3.66}$$

where $\widetilde{x}_\delta$ and $\widetilde{x}_e$ represent two components of the state of the error dynamics; namely, one due to the faulty signal δ and one arising from the exogenous disturbance e, respectively. In Eq. (3.66), signal δ is filtered and differentiated.

For the purpose of illustration, consider a typical abrupt signal: the unit step. Also, suppose the smooth exogenous disturbance is a sinusoidal function. Both signals are shown in Fig. 3.18(a). With an appropriate selection of parameters τ_f and L_o, one can expect the component of the state of the error dynamics due to faulty signal δ to dominate, amplitude-wise, the other component of the state of the error dynamics, $\widetilde{x}_e$. This is suggested by the signals shown in Fig. 3.18(b), and by the root mean square of the state of the error dynamics $\|\widetilde{x}\|_\tau$ defined in Eq. (3.68) and

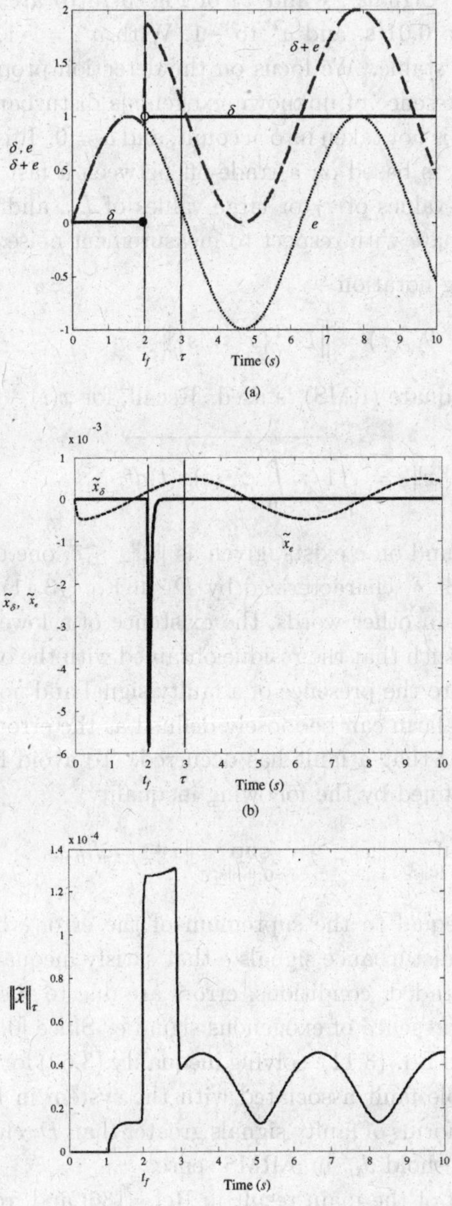

Fig. 3.18 Observer with compensation of disturbances. (a) Profiles of a faulty signal (solid line), an exogenous disturbance (dotted line), and the combination of both (dashed line). (b) Error signals $\widetilde{x}_\delta$ and $\widetilde{x}_e$. (c) Root mean square of the state of the error dynamics $\|\widetilde{x}\|_\tau$.

shown in Fig. 3.18(c). Signals $\widetilde{x}_\delta$ and $\widetilde{x}_e$ of Fig. 3.18(b) are obtained by setting L_o to 20, τ_f to 0.01 s, and a^* to -1. With $a^* = -1$, the system given by Eq. (3.64) is stable. We focus on the detection property of Eqs. (3.64) and (3.65) in presence of unknown exogenous disturbances. Hence, parameter uncertainty is not taken into account, and $\widetilde{a} = 0$. It is to be noted that tuning τ_f and L_o is based on a trade-off between a fast detection of δ obtained with small values of τ_f or large values of L_o, and maintaining robustness of the estimate with respect to measurement noise.

Adopt the following notation

$$\Lambda_x(\tau) = \left\| L^{-1}\{x(s)/s\} \right\|_\tau , \tag{3.67}$$

where the root mean square (RMS) is used. Recall, for $x(t)$,

$$\|x\|_\tau = \sqrt{1/\tau \int_0^\tau x^\tau(t)x(t)dt}. \tag{3.68}$$

When an upper bound on e exists, given as $\|e\|_\tau \leq \bar{e}$, one can compute a lower detectable fault δ, characterized by D_i in Eq. (3.11), and a corresponding threshold. In other words, the existence of a lower detectable fault means that D_i is such that the residue obtained with the observer rises above a threshold due to the presence of a faulty signal and not because of a false alarm. A false alarm can be loosely defined as the erroneous acceptance of the hypothesis that a fault has occurred. To avoid false alarms, the threshold J_{th} is defined by the following inequality

$$\inf_{\delta \in \Omega_\delta, \|e\| \leq \bar{e}} \|\widetilde{x}\|_\tau > \sup_{\delta = 0, \|e\| \leq \bar{e}} \|\widetilde{x}\|_\tau = J_{th}. \tag{3.69}$$

The threshold J_{th} is equal to the supremum of the error when no fault occurs, $\delta = 0$, over all disturbance signals e that satisfy inequality $\|e\| \leq \bar{e}$. In normal, or non-degraded, conditions, errors are due to the parametric uncertainties and the presence of exogenous signal e. Since $[0,0,0]^T \notin \Omega_\delta$, by construction of Ω_δ in Eq. (3.11), solving inequality (3.69) for D_i provides the minimum detectable fault associated with the system in Eqs. (3.63); that is, the Euclidean norms of faulty signals greater than D_i entail residues $\widetilde{x}$ greater than the threshold J_{th} in a RMS sense.

Following the proof of the main result in Ref. [136] and from the definition of $H(s)$ in Eq. (3.66), inequality (3.69) is satisfied if

$$\inf_{\delta \in \Omega_\delta} \|H(s)\delta(s)\|_\tau > 2 \sup_{\|e\| \leq \bar{e}} \|H(s)e(s)\|_\tau . \tag{3.70}$$

To obtain such sufficient, although somewhat conservative, condition, it should be mentioned that

$$\inf_{\delta \in \Omega_\delta, \|e\| \leq \overline{e}} \|\widetilde{x}\|_\tau > \inf_{\delta \in \Omega_\delta} \|H(s)\delta(s)\|_\tau .$$

From the definitions of the root mean square and D_i in Eq. (3.11) with $i = 1$, $\delta \in \Omega_\delta$ implies that the left-hand side of inequality (3.69) can be bounded from below by $|D_i|\Lambda_H(\tau - t_f)$. Inequality (3.70) thus yields the following sufficient condition on the minimum detectable fault:

$$|D| \geq \frac{2\overline{e}\,\|H(s)\|_\infty}{\Lambda_H(\tau - t_f)}. \tag{3.71}$$

From inequality (3.69), the threshold is thus defined as

$$J_{th} = 2\overline{e}\,\|H(s)\|_\infty .$$

For the system given by Eqs. (3.63), assuming there are no parametric uncertainties ($\widetilde{a} = 0$), $\Lambda_H(\tau - t_f)$ can be expressed in closed form as

$$\begin{aligned} \Lambda_H(\tau - t_f) &= \sqrt{1/(\tau - t_f) \int_0^{\tau - t_f} \left[L^{-1}\{H(s)/s\}\right]^T L^{-1}\{H(s)/s\} dt} \\ &= \tfrac{a^2 t_f^2}{\tau(t_f(a+L_o)-1)^2} \left[e^{-2(a+L_o)(\tau - t_f)} + e^{-2(\tau - t_f)} - 2e^{-(a+L_o+1/t_f)(\tau - t_f)} - 3\right]^2 . \end{aligned} \tag{3.72}$$

The concept of observer with disturbance attenuation, illustrated with the first-order linear, time-invariant system of Eqs. (3.63), forms the basis of DAFD. First, an observer is devised. The observer generates a residue which is sensitive to the faulty signal despite the unknown exogenous disturbances. A filtered derivative of the appropriate signals is used to reconstruct signals of the type $\widehat{\delta} + \widehat{e}$. Second, a threshold is calculated. The selection of the threshold is critical to enable the detection of abrupt faults while avoiding false alarms.

3.3.4 *Synthesis of DAFD*

The design of DAFD relies on the model given by Eqs. (3.60). The design process is complicated by the fact that some unknown exogenous disturbances excite the error dynamics of the observer. To understand the origin of the unknown disturbance, consider Fig. 3.19. The figure shows a string-type formation of three vehicles, labeled as k, i, and j. Information flows from j to i, and from i to k. UAV k monitors vehicle i by using the model presented in Eqs. (3.60), which is assumed to represent the closed-loop dynamics of i. UAV i receives information on the positions x_j and y_j of j.

Vehicle j is therefore in N_i, by virtue of Definition 3.1. For this example, based on the limited sensing capabilities of vehicle k, UAV j is not in N_k. It is well known that the effectiveness of an observer depends on the availability of input-output pairs of the system being estimated. If x_j and y_j were available to vehicle k, the observer onboard vehicle k, which monitors vehicle i, would use such signals with the block diagram shown in Fig. 3.19(b). In such case, the design of the observer would be straightforward. Here, x_j and y_j are not available to the observer onboard UAV k. The situation is illustrated in Fig. 3.19(c). Therefore, for the computation of the residue, the observer is constrained to rely on the position of vehicle k and on the relative position between vehicles k and i, namely variables x_{ki} and y_{ki}. This lack of information on vehicle j has an impact on the observer error dynamics: it takes the form of an unknown exogenous disturbance. As is done in Section 3.3.3, the DAFD observer is designed such that the impact of the exogenous disturbance on the residue is attenuated.

Considering that signal x^*_{ij} in (3.60) is available to k, the following observer is proposed

$$\dot{\widehat{q}}_i = A_{F,i}\widehat{q}_i + B^*_i \begin{bmatrix} h_i \sum_{j\in N_i} k_i(\widehat{x}_j - x^*_{ij}) \\ h_i \sum_{j\in N_i} k_i(\widehat{y}_j - y^*_{ij}) \\ z_t \end{bmatrix} + B_{F,i}C_i q_i, \tag{3.73}$$

$$r_i = \begin{bmatrix} C_i & -C_{F,i} \end{bmatrix} \begin{bmatrix} q_i \\ \widehat{q}_i \end{bmatrix},$$

where

$$\begin{aligned} B^*_i \begin{bmatrix} h_i \sum_{j\in N_i} k_i\widehat{x}_j \\ h_i \sum_{j\in N_i} k_i\widehat{y}_j \\ 0 \end{bmatrix} &= L^{-1}\left\{\frac{B^*_i \widehat{v}_{i1}(s)}{\tau_f s+1}\right\} \\ &= L^{-1}\left\{\frac{\overline{q}_i(s) - A^*_i \overline{q}_i(s) + B^*_i v_{i2}(s)}{\tau_f s+1}\right\} \end{aligned} \tag{3.74}$$

and

$$\overline{q}_i(s) = \begin{bmatrix} q_{i1}(s) \\ \frac{s x_i(s)}{\tau_d s+1} \\ q_{i3}(s) \\ \frac{s y_i(s)}{\tau_d s+1} \\ q_{i5}(s) \\ q_{i6}(s) \end{bmatrix}. \tag{3.75}$$

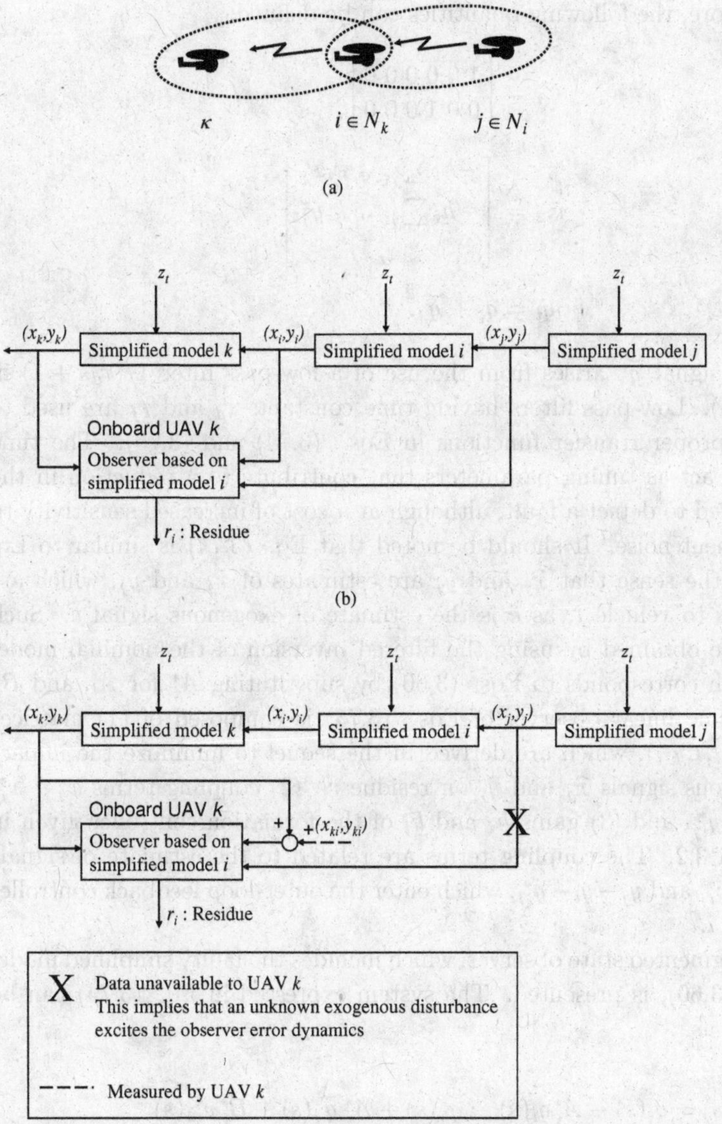

Fig. 3.19 Example of three-vehicle formation and observer. (a) String formation. (b) Ideally, the observer onboard vehicle k uses all input-output pairs of the model of vehicle i. (c) Actual observer relies on position of vehicle k and on relative positions between k and i.

Furthermore, the following quantities can be defined:

$$C_i = \begin{bmatrix} 1\,0\,0\,0\,0\,0 \\ 0\,0\,1\,0\,0\,0 \end{bmatrix},$$

$$v_{i2} = \begin{bmatrix} -h_i \sum_{j\in N_i} k_i x_j^* \\ -h_i \sum_{j\in N_i} k_i y_j^* \\ z_t \end{bmatrix},$$

$$q_i = \overline{q}_i + \breve{q}_i.$$

Error signal $\breve{q}_i$ arises from the use of a low-pass filter $1/(\tau_d s + 1)$ in Eq. (3.75). Low-pass filters having time constants τ_d and τ_f are used to obtain biproper transfer functions in Eqs. (3.74) and (3.75). The time constants act as tuning parameters that contribute to a reduction in the time needed to detect a fault, although at a cost of increased sensitivity to measurement noise. It should be noted that Eq. (3.74) is similar to Eq. (3.65) in the sense that $\widehat{x}_j$ and $\widehat{y}_j$ are estimates of x_j and y_j, which are exogenous to vehicle i, as $\hat{e}$ is the estimate of exogenous signal e. Such signals are obtained by using the filtered inversion of the nominal model of i, which corresponds to Eqs. (3.60) by substituting A_i^* for A_i, and B_i^* for B_i. The linear observer of Eqs. (3.73) is composed of (1) matrices $(A_{F,i}, B_{F,i}, C_{F,i})$, which are derived in the sequel to minimize the impact of exogenous signals $\widehat{x}_j$ and $\widehat{y}_j$ on residue r_i, (2) coupling terms $\widehat{x}_j - x_{ij}^*$ and $\widehat{y}_j - y_{ij}^*$, and (3) gains h_i and k_i of the formation controller given in Definition 3.2. The coupling terms are related to the estimate of signals $x_j - x_i - x_{ij}^*$ and $y_j - y_i - y_{ij}^*$, which enter the outer-loop feedback controller of vehicle i.

An augmented state observer, which includes the faulty simplified model in Eqs. (3.60), is presented. The system expressed in Eq. (3.74) can be expanded as

$$B_i^* \widehat{v}_{i1}(s) = \dot{q}_i(s) - A_i^* q_i(s) - \dot{\breve{q}}_i(s) + A_i^* \breve{q}_i(s) + B_i^* v_{i2}(s)$$

$$= \widetilde{A}_i q_i(s) + B_i(v_{i1}(s) - v_{i2}(s)) + \delta_i(s) - \dot{\breve{q}}_i(s) + A_i^* \breve{q}_i(s) + B_i^* v_{i2}(s) \tag{3.76}$$

where the first equality is obtained by substituting $\overline{q}_i + \breve{q}_i$ for q_i in Eq. (3.74). The second equality is obtained by replacing q_i by the simplified model given in Eqs. (3.60) to which is added the faulty signal δ_i. Then, letting

$$\begin{aligned}
&\alpha_{i2}(s) = \tfrac{B_i v_{i1}(s) - \widetilde{B}_i v_{i2}(s) - \breve{q}_i(s) + A_i^* \breve{q}_i(s)}{\tau_f s + 1}, \\
&A_i'(\alpha) = \begin{bmatrix} -1/\tau_f & 0 \\ 0 & A_i(\alpha) \end{bmatrix}, \widetilde{A}_i'(\alpha) = \begin{bmatrix} \widetilde{A}_i(\alpha) \; 0 \end{bmatrix}, \\
&B_i'(\alpha) = \begin{bmatrix} 0 \\ B_i(\alpha) \end{bmatrix}, C_i' = \begin{bmatrix} 0 \; C_i \end{bmatrix}, Q_i = \begin{bmatrix} \varsigma_{1i} \\ q_i \end{bmatrix}, \\
&\varsigma_{2i} = (\delta_i - \varsigma_{2i})/\tau_f, \; \Delta_i = \begin{bmatrix} \delta_i \; \varsigma_{2i} \end{bmatrix}^T,
\end{aligned} \tag{3.77}$$

where state-space variables ς_{1i} and ς_{2i} follow from the realization of the filter $1/(\tau_f s+1)$ used in Eq. (3.74) and from the filtering of δ_i, respectively, yields the following augmented state observer

$$\begin{aligned}
\begin{bmatrix} \dot{Q}_i \\ \dot{\widehat{q}}_i \end{bmatrix} &= \underbrace{\begin{bmatrix} A_i'(\alpha) & 0 \\ B_{F,i} C_i' + \widetilde{A}_i'(\alpha) & A_{F,i} \end{bmatrix}}_{=\mathrm{A}_i(\alpha)} \begin{bmatrix} Q_i \\ \widehat{q}_i \end{bmatrix} \\
&+ \underbrace{\begin{bmatrix} B_i'(\alpha) & 0 \\ 0 & I \end{bmatrix}}_{=\mathrm{B}_i(\alpha)} \begin{bmatrix} v_{i1} - v_{i2} \\ \alpha_{i2} \end{bmatrix} + \begin{bmatrix} \begin{bmatrix} 0 \\ 1 \end{bmatrix} & 0 \\ 0 & 1 \end{bmatrix} \Delta_i, \\
z &= \begin{bmatrix} C_i' & -C_{F,i} \end{bmatrix} \begin{bmatrix} Q_i \\ \widehat{q}_i \end{bmatrix}.
\end{aligned} \tag{3.78}$$

In Eqs. (3.78), the equation for $d\widehat{q}_i/dt$ is obtained (1) by substituting the right-hand side of the second equality in Eq. (3.76) for $B_i^* \widehat{v}_{i1}(s)$ in Eqs. (3.73), (2) by using $B_i^* = B_i - \widetilde{B}_i$, and (3) by noticing that $B_i v_{i2}(s)$ vanishes. The equation for the derivative of Q_i comes from Eqs. (3.60) augmented with ς_{1i}.

It is interesting to note that a Luenberger-type observer yields a residue z that can be expressed, similarly to $H(s)\delta(s)$ in Eq. (3.66), as a function of a low-pass filtered derivative of the faulty signal. Setting $A_{F,i}$ to A_i^*, and $C_{F,i}$ to C_i leads to

$$\begin{aligned}
z(s) &= \underbrace{C_i'(sI - A_i^{*\prime})^{-1} \widetilde{A}_i}_{=H_1(s)} Q_i(s) \\
&\underbrace{+ C_i'(sI - A_i^{*\prime})^{-1} B_i'(\alpha)(v_{i1}(s) - v_{i2}(s)) - C_i(sI - A_i^*)^{-1} \alpha_{i2}(s)}_{=H_2(s)} \\
&\underbrace{- C_i(sI - A_i^*)^{-1} \frac{\tau_f s}{\tau_f s + 1}}_{=G_i(s)} \delta(s),
\end{aligned} \tag{3.79}$$

where $A_i^{*\prime} = A_i^{\prime}(\alpha^*)$. The contribution of $G_i\delta(s)$ to $z(s)$ in Eq. (3.79) is similar to that of $H(s)\delta(s)$ to $\widetilde{x}(s)$ in Eq. (3.66).

The computation of observer matrices $A_{F,i}$, $B_{F,i}$, and $C_{F,i}$ is obtained by solving the L$_2$-gain minimization problem defined as

$$\min_{A_{F,i},B_{F,i},C_{F,i}} \{\gamma; \|T_{\alpha_{i2}z_i}\|_\infty < \gamma\}, \tag{3.80}$$

where

$$T_{\alpha_{i2}z_i} = \left(A_i, B_i, \left[\, C_i^{\prime} \; -C_{F,i} \,\right], 0\right).$$

A sufficient condition to satisfy the minimization constraint

$$\|T_{\alpha_{i2}z_i}\|_\infty < \gamma$$

is given by the following matrix inequality

$$\begin{bmatrix} P\mathrm{A}_i(\alpha) + \mathrm{A}_i^T(\alpha)P & P\mathrm{B}_i(\alpha) & \left[\, C_i^{\prime} \; -C_{F,i} \,\right]^T \\ \mathrm{B}_i^T(\alpha)P & -\gamma^2 I & 0 \\ \left[\, C_i^{\prime} \; -C_{F,i} \,\right] & 0 & -I \end{bmatrix} < 0. \tag{3.81}$$

Inequality (3.81) depends on matrix products $PA_i(\alpha)$ and $PB_i(\alpha)$, which contain products of observer matrices and elements of decision matrix P. Following the procedure of Ref. [137], it can be shown that inequality (3.81) is equivalent, for all $j \in \{1, ..., s\}$, to a LMI that can be efficiently solved by software, such as the LMI toolbox from The MathWorks [108], and Scilab [124], to name a few. To pursue the development of DAFD, let us adopt the following notation: symbol $*$ stands for the entries of a symmetrical matrix,

$$\Theta = [I \; 0]^T, C_j = diag(C_i^{\prime}, I),$$

and

$$\Im = [I \; I].$$

The robustness property of DAFD is stated in Proposition 3.1.

Proposition 3.1. *Robustness of DAFD* [15]. *The observer given by Eqs. (3.73)-(3.75) and characterized by matrices $A_{F,i}$, $B_{F,i}$, and $C_{F,i}$ solves the L$_2$-gain minimization problem (3.80) whenever there exists $\mu_i > 0$ such* *that the following LMI*

$$
\begin{bmatrix}
e_{11} & e_{12} & \widehat{V}_i^T \Theta B_{ij} & 0 & \mu_i \widehat{V}_i^T \\
* & -\widehat{X}_i & 0 & e_{24} & 0 \\
* & * & -\gamma^2 I & 0 & 0 \\
* & * & * & -I & 0 \\
* & * & * & * & -\widehat{X}_i
\end{bmatrix} < 0,
$$

$$
\begin{aligned}
&\mu_i > 0, \\
&\widehat{X}_i > 0, \\
&e_{11} = -\mu_i(\widehat{V}_i + \widehat{V}_i^T), \\
&e_{12} = \widehat{V}_i^T \Theta A_{ij} \Theta^T + \Im^T \widehat{K}_i C_j + \Im^T \widetilde{A}_{ij} \Theta^T + \widehat{X}_i^T, \\
&e_{24} = \begin{bmatrix} C_i'^T \\ -\widehat{C}_{F,i}^T \end{bmatrix},
\end{aligned} \tag{3.82}
$$

is feasible, for all A_{ij} *and* B_{ij} *with* $j \in \{1, ..., s\}$, *in*

$$
\widehat{V}_i = \begin{bmatrix} \widehat{V}_{1,i} & \widehat{V}_{2,i} \\ \widehat{V}_{3,i} & \widehat{V}_{3,i} \end{bmatrix},
$$

and for $\widehat{X}_i$, $C_{F,i}$, *and* $\widehat{K}_i$. *The observer matrices are expressed as*

$$
A_{F,i} = \widehat{A}_{F,i} \widehat{V}_{3,i}^{-T},
$$

$$
B_{F,i} = \widehat{B}_{F,i},
$$

and

$$
C_{F,i} = \widehat{C}_{F,i} \widehat{V}_{3,i}^{-T}.
$$

The following result is taken from Ref. [137] and is crucial for the proof of Proposition 3.1. It is useful to obtain the linearization of matrix-inequality-based problems, such as L_2-gain minimization, where the Lyapunov variable is multiplied by a design variable.

Theorem 3.2. [137]. *LMI*

$$
\begin{aligned}
&X > 0, \\
&\begin{bmatrix} A^T X + XA & XB & C^T \\ B^T X & Q_{11} & Q_{12} \\ C & Q_{12}^T & Q_{22} \end{bmatrix} < 0,
\end{aligned} \tag{3.83}
$$

is feasible in the decision variable X if and only if there is a scalar $\mu > 0$ such that the following LMI

$$\begin{bmatrix} -\mu(V+V^T) & V^TA+X & V^TB & 0 & \mu V^T \\ A^TV+X & -X & 0 & C^T & 0 \\ B^TV & 0 & Q_{11} & Q_{12} & 0 \\ 0 & C & Q_{12}^T & Q_{22} & 0 \\ \mu V & 0 & 0 & 0 & -X \end{bmatrix} < 0 \tag{3.84}$$

is feasible in the decision variables X and V.

Several LMIs, such as (3.84), are provided in Ref. [137]. We retain the LMI that is useful for the design of a robust observer. From inequalities (3.81) and (3.83), one has:

$$C = [\,C_i' \; -C_{F,i}\,],$$

$$Q_{12} = 0, Q_{11} = -\gamma^2 I, Q_{22} = -I.$$

Proof. The proof of Proposition 3.1 is similar to that of Theorem 5 in Ref. [137] applied to Eqs. (3.78) and (3.81). The extra coupling term $\widetilde{A}_i'(\alpha)$ in $\mathrm{A}_i(\alpha)$, which is not present in Theorem 5 of Ref. [137], yields term $\Im^T \widetilde{A}_{ij}\Theta^T$ in the $(1,2)$ entry of the matrix in LMI (3.82). This product comes from $\Gamma\widetilde{A}_{ij}\Theta^T$, which is part of $\mathrm{A}_i(\alpha)$ given as

$$\mathrm{A}_i(\alpha) = \sum_{j=1}^{s} \alpha_{ij}\left(\Theta A_{ij}\Theta^T + \Gamma K_i C_j + \Gamma\widetilde{A}_{ij}\Theta^T\right), \tag{3.85}$$

where

$$\Gamma = [0 \; I]^T,$$

and

$$K_i = [\,B_{F,i} \; A_{F,i}\,].$$

To obtain the LMI in (3.82), the congruence transform $diag(\Pi_{V_i}, \Pi_{V_i}, 1, 1, \Pi_{V_i})$, where

$$\Pi_{V_i} = \begin{bmatrix} I & 0 \\ 0 & V_{22,i}^{-1}V_{21,i} \end{bmatrix},$$

$$V_i = \begin{bmatrix} V_{11,i} & V_{12,i} \\ V_{21,i} & V_{22,i} \end{bmatrix} = \begin{bmatrix} \widehat{V}_{1,i} & \widehat{V}_{2,i}\widehat{V}_{3,i}^{-T} \\ I & \widehat{V}_{3,i}^{-T} \end{bmatrix},$$

is first applied to LMI (3.84) in Theorem 3.2. Note that LMI (3.84) is applied to matrix inequality (3.81) and uses Eq. (3.85). Then, the following useful relationships are employed:

$$\Pi_{V_i}\Theta A_{ij}^T\Theta^T V_i\Pi_{V_i} = \Theta A_{ij}^T\Theta^T\widehat{V}_i,$$

$$\Pi_{V_i}^T C_j^T K_i^T\Gamma^T V_i\Pi_{V_i} = C_j^T\widehat{K}_i^T\Im,$$

and

$$B_{ij}^T\Theta^T V_i\Pi_{V_i} = B_{ij}^T\Theta^T\widehat{V}_i.$$

As already noticed, the congruence transform yields the extra term $\Pi_{V_i}^T\Theta\widetilde{A}_{ij}\Gamma^T V_i\Pi_{V_i}$, which can be shown to be equal to $\Theta\widetilde{A}_{ij}^T\Im$ by noting that

$$\Gamma^T V_i\Pi_{V_i} = V_{21,i}\Im = \Im$$

and

$$\Pi_{V_i}^T\Theta = \Theta.$$

□

3.3.5 *Threshold selection*

The DAFD system determines if a change in the signal of interest constitutes a fault by comparing the RMS of the residue with a threshold. Faulty behavior is usually declared when the RMS of the residue is greater than or equal to the threshold. The minimum detectable fault and the corresponding threshold are expected to be functions of leader-follower relative distances, and thus should be selected carefully. Indeed, the tracking error of each vehicle increases in norm with the leader-to-follower relative distance. In fact, constraining tracking errors within small bounds is similar to forcing the entire formation to behave like a rigid body, which could entail prohibitively large control signals for vehicles significantly remote from the leader [105, 115]. In the long term, large control inputs may impair the actuators.

As mentioned for the illustrative case of Sec. 3.3.3, and expressed in inequality (3.69), to avoid false alarms the threshold should be defined such that

$$\inf_{\delta\in\Omega,Q,v_{i1},v_{i2},\alpha_{i2}} \|z_i\|_\tau > \inf_{\delta=0,Q,v_{i1},v_{i2},\alpha_{i2}} \|z_i\|_\tau\,. \tag{3.86}$$

From Ref. [136], inequality (3.86) is satisfied if

$$\inf_{\delta\in\Omega} \|G_i(s)f_i(s)\|_\tau > 2 \cdot \sup_{Q,v_{i1},v_{i2},\alpha_{i2}} \left\| H_1(s)\widetilde{A}_i(s)Q_i(s) + H_2(s) \right\|_\tau . \quad (3.87)$$

The minimum detectable fault associated with vehicle i can be characterized by the following lower bound on D_i, introduced in Eq. (3.11),

$$|D_i| \geq 2 \cdot \frac{\sup_{Q,v_{i1},v_{i2},\alpha_{i2}} \left(\left\| H_1(s)\widetilde{A}_i(s) \right\|_\infty \|Q_i(s)\|_\tau + \|H_2(s)\|_\tau \right)}{\underline{\sigma}(\Lambda_{G_i(s)}(\tau - t_f))} \quad (3.88)$$

where $\underline{\sigma}$ denotes the minimum singular value of its argument. Since $G_i(s)$ is a matrix, $\Lambda_{G_i(s)}(\tau - t_f)$ in inequality (3.88) is defined as the following matrix operator

$$\sqrt{\frac{1}{\tau}\int_0^T \left[L^{-1}\{\frac{G_i(s)}{s}\} \right]^T L^{-1}\{\frac{G_i(s)}{s}\} dt.}$$

The Euclidean norm of position and speed tracking errors of vehicle i is an increasing function of the relative distance ρ_{i0}^* between the leader and follower i. To see this, recall that (1) the radius R_{B_d} of ball B_d, introduced in Theorem 3.1, is bounded from above by

$$\sup_{\theta_0,\dot{\theta}_0,\ddot{\theta}_0,\ddot{q}^*} \overline{\sigma}_\Omega \left\| d(\theta_0, \dot{\theta}_0, \ddot{\theta}_0, \ddot{q}^*) \right\|,$$

where $\overline{\sigma}_\Omega$ denotes the largest singular value of Ω, and (2) by virtue of inequality (3.57), $\left\| d(\theta_0, \dot{\theta}_0, \ddot{\theta}_0, \ddot{q}^*) \right\|$ is bounded by an increasing function of ρ_{i0}^*. From such facts, the following bound can be derived

$$\|Q_i\|_\tau \leq \|Q_i\|_2 \leq (a_{i1} + \varphi_{i1}(\rho_{i0}^*))/2,$$

where $a_{i1} > 0$, and φ_{i1} is an increasing function. Thus, a sufficient condition for the detection of a fault δ_i can be expressed as

$$|D_i| \geq \frac{J_{th,i}}{\underline{\sigma}(\Lambda_{G_i(s)}(\tau - t_f))}, \quad (3.89)$$

where

$$J_{th,i} \geq (a_{i1} + \varphi_{i1}(\rho_{i0}^*)) \left\| H_1(s)\widetilde{A}_i(s) \right\|_\infty + \|H_2(s)\|_\tau . \quad (3.90)$$

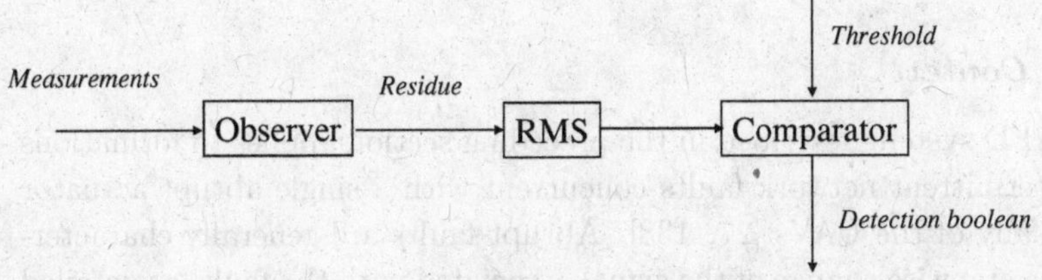

Fig. 3.20 Schematics of DAFD function.

Inequality (3.89) shows that false alarms can be avoided by increasing $J_{th,i}$ as a function of ρ_{i0}^*. However, an increase in $J_{th,i}$ may impede detection of fault for the vehicles whose distances from the leader are significant. There is therefore a trade-off in the selection of the threshold.

3.3.6 *DAFD design algorithm*

The complete DAFD function is illustrated schematically in Fig. 3.20. Such structure should be implemented onboard each vehicle. If a fault is detected, the detection boolean is set to 1, otherwise it is 0. The steps in the design of the DAFD function are given as follows.

Step 1. Obtain a simplified model of the neighboring vehicles. A model is given by Eqs. (3.60).

Step 2. Design an observer defined by Eqs. (3.73)-(3.75). Obtain the observer matrices $A_{F,i}$, $B_{F,i}$, and $C_{F,i}$ by solving the L$_2$-gain minimization problem (3.80). Tune τ_d and τ_f by means of numerical simulations.

Step 3. Select a threshold $J_{th,i}$ satisfying inequality (3.90) to avoid false alarms. Tuning the threshold may be done by means of numerical simulations. It is important to note that a threshold may bear a single value or may come from a set of values whose selection depends on the position of a follower vehicle in a formation and its distance from the leader.

DAFD relies on the information obtained by the onboard sensors. If DAFD detects a fault on a neighboring UAV, an adaptation of the command takes place (Sec. 3.5). If no UAV is within range, DAFD cannot carry out its function.

3.4 Signal-Based Decentralized Non-Abrupt Fault Detector

3.4.1 *Context*

The DAFD system described in the preceding section applies to formations with intermittent network faults concurrent with a single abrupt actuator fault in any of the UAVs [27, 138]. Abrupt faults are generally characterized by a stepwise change of the signal associated with the fault, as detailed in Sec. 3.1.3. However, decentralized abrupt fault detection schemes are insensitive to certain types of non-abrupt actuator faults, including lock-in-place faults [93]. To circumvent this limitation, and to devise a complementary functionality, we present a non-abrupt fault detection scheme based on signal analysis rather than on state-space models. We call this monitoring function a decentralized non-abrupt fault detector (DNaFD). Assuming network communications are operational and a minimum number of sensors are healthy, embedding DNaFD onboard every vehicle in the formation enables any given vehicle (labeled i) to monitor its neighbors within range (vehicles j_i), and to detect faulty behaviors. DNaFD requires a minimum of two signals. The heading angle trajectory of vehicle i is compared to the network-communicated heading angle trajectory of at least one vehicle that is a neighbor of j_i. By neighbor of j_i, we mean a vehicle whose separation from j_i is a signal employed by the formation control law. The level of correlation between these two signals allows discrimination between the transients resulting from the motion of a healthy vehicle from the transients associated with a faulty behavior.

DNaFD is shown to effectively detect faults on neighboring vehicles for the following scenario: A vehicle is faced with a non-abrupt fault on one of its actuators. The network is assumed to be operational, although the information emanating from the faulty vehicle j_i is taken as unreliable. Such a situation may arise, for example, when a collision causes damages to sensors and actuators on vehicle j_i, resulting in uncompensated vehicle drift, erroneous measurements, and poor state estimates. DNaFD relies on fault propagation from j_i to i due to the dynamic couplings inherent in distributed formation control.

The basic principle of DNaFD is illustrated in Fig. 3.21. Information is obtained via sensors and the communication network. Follower 2 is at fault, and is being monitored by followers 3 and 4. Meanwhile, follower 1 is being monitored by followers 2 and 3. It is clear that the control system onboard follower 3 can determine whether a fault has occurred by

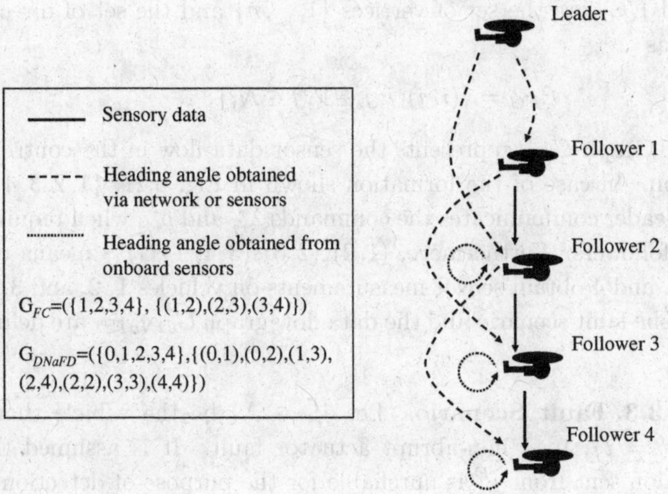

Fig. 3.21 Information flow for formation control (G_{FC}) and DNaFD (G_{DNaFD}).

comparing signals from followers 1 and 2. In the figure, the signal of interest for the purpose of comparison is UAV heading. Graph G_{FC} represents sensor data used for the control of the formation and for transmission of geometry commands. Data flow graph G_{DNaFD} represents information flow relevant to DNaFD, which is shown for follower vehicles 2 to 4 in Fig. 3.21. We thus assume that the current leader is fault free, and so follower 1 does not carry out DNaFD on its immediate predecessor, which is the leader vehicle. In fact, follower 1 cannot carry out DNaFD on the leader vehicle due to the lack of information. Networking vehicle information and leveraging sensor measurements are expected to be enablers of non-abrupt fault detection provided the processing of signals is carefully addressed, as discussed below.

3.4.2 *Networked information and coupling effects*

For the purpose of synthesizing the detector, let us define $j_i \in \mathcal{N}_i$ and $k_{j,i} \in \mathcal{N}_{j_i}$ as the neighbors of vehicles i and j_i, respectively. Formation geometry commands, characterized by angles λ^*_{ij} and relative distances ρ^*_{ij}, are transmitted from one node to another according to graph

$$G_{FC} = (\mathcal{V}, E_{FC}),$$

where $\mathcal{V}$ and E_{FC} are the set of vertices $\{1, ..., n\}$ and the set of directed edges given as

$$E_{FC} = \{(i,j); i,j \in \mathcal{V}, j \in \mathcal{N}_i\},$$

respectively. Graph G_{FC} represents the sensor data flow in the control of the formation. In case of the formation shown in Fig. 3.21, $\{1,2,3,4\} \in G_{FC}$ as the leader communicates the commands λ^*_{ij} and ρ^*_{ij}, when required, to the four followers. Furthermore, $(1,2), (2,3), (3,4) \in G_{FC}$ means that vehicles 2, 3, and 4 obtain sensor measurements on vehicles 1, 2, and 3, respectively. The fault scenario and the data flow graph G_{DNaFD} are defined as follows.

Definition 3.3. Fault Scenario. Let $j_i^F \in \mathcal{N}_i$ be the vehicle that is subject, at $t = t_f$, to a non-abrupt actuator fault. It is assumed that the information sent from j_i^F is unreliable for the purpose of detection by vehicle i. Furthermore, it is assumed that two or more non-abrupt faults cannot occur within $[t_f, t_f + \tau)$, where τ stands for the time interval needed to collect data and to perform hypothesis testing. Denote T_s as the time period between updates on the onboard sensor signals.

Definition 3.4. Data Flow Graph. DNaFD uses data flow graph

$$G_{DNaFD} = (\mathcal{V}, E_{DNaFD}),$$

where the set of vertices $\mathcal{V} = \{1, ..., n\}$ is defined in Sec. 3.2.2. E_{DNaFD} is the set of directed edges (i,i), and $(i, k_{j,i})$, where $i, k_{j,i} \in \mathcal{V}$, and $k_{j,i} \in \mathcal{N}_{j_i}$. The edges of G_{DNaFD} represent information flow relevant to DNaFD; that is, $k_{j,i}$ sends its estimated heading angle to i by means of the communication network whose sample period is $T_n = pT_s$, $p \in \mathbb{N}$. Thus, the rate at which sensor measurements are updated is assumed faster than that for the communication network. We assume the information sent by j to i is not taken into account if j is faulty. For example, vehicle j may be drifting because of a sensor fault. In such case, heading angle estimates or measurements may be erroneous, and neighboring vehicles should not rely on such signals.

In a few words, DNaFD relies on the fact that each follower node of G_{DNaFD} has indegree greater than or equal to two to perform a fault-no-fault hypothesis test on the available signals. To describe the DNaFD function in detail, we focus on one aerial platform : the ALTAV. Numerous simulations of the nonlinear dynamics of the ALTAV indicate that the heading angle is relatively sensitive to non-abrupt actuator faults; therefore making such signal relevant for the operation of DNaFD. This may be explained by the fact that an asymmetric use of antagonistic actuators, intentional or not, impacts on the speed vector orientation. Typical heading angles obtained with the ALTAVs are shown in Fig. 3.22. Consider a formation, or sub-formation, of three vehicles. Vehicle i follows vehicle j_i^F, which in turns follows vehicle $k_{j,i}$. A non-abrupt fault occurs on an actuator of vehicle j_i^F at time t_f. A follower trajectory is similar to that of its immediate predecessor, albeit with a time delay, from the dynamic coupling effects induced by formation control; namely, the coupling of G_{FC} as given in Eqs. (3.58). A fault occurring at t_f onboard vehicle j_i^F entails either a new steady-state heading angle trajectory or a drifting trajectory, as depicted in Fig. 3.22(b) by curves (A) and (B), respectively. The DNaFD function implemented onboard vehicle i consists of a discrete-time estimator, with sampling period T_s, and a statistical test, whose decision function is given as

$$g : \Omega^N \rightarrow \{H_0, H_1\}$$

with Ω being the sample space and

$$g(t_k, \mathcal{Y}_{1,N}) = \begin{cases} H_0 \text{ if } \mathcal{T}(t_k, \mathcal{Y}_i) = 0, \\ H_1 \text{ if } \mathcal{T}(t_k, \mathcal{Y}_i) = 1. \end{cases} \tag{3.91}$$

H_0 and H_1 denote normal and faulty behaviors, respectively. Function $\mathcal{T}$ is defined over

$$\mathcal{Y}_i = \{\mathcal{Y}_{1,N_{k_{j,i}}}(k_{j,i}), \mathcal{Y}_{1,N_i}(i)\}.$$

Sequences of heading angles

$$\mathcal{Y}_{1,N_{k_{j,i}}}(k_{j,i}) = \{\widehat{\psi}_{k_{j,i}}(T_s), ..., \widehat{\psi}_{k_{j,i}}(N_{k_{j,i}}T_s)\}$$

and

$$\mathcal{Y}_{1,N_i}(i) = \{\widehat{\psi}_i(T_s), ..., \widehat{\psi}_i(N_iT_s)\}$$

are stored onboard vehicle i. The heading angle $\widehat{\psi}_i$ is obtained either from measurement or from a state estimator. Sequence $\mathcal{Y}_{1,N_{k_{j,i}}}$ contains information on the heading angles of vehicle $k_{j,i}$ whereas sequence $\mathcal{Y}_{1,N_i}$ pertains to heading angles of vehicle i. Two sequences are required to carry out the comparison. There may be more than a single pair of sequences available, depending on the information flow. In such case, additional information is available for DNaFD, although the problem of fault detection becomes more complex. In this section, a single pair of sequences is considered, assuming

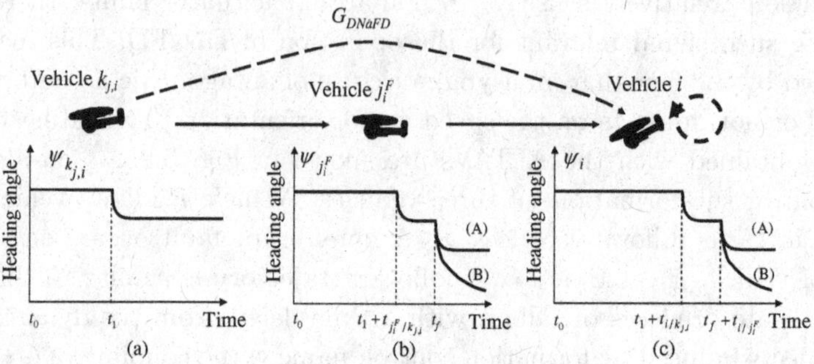

Fig. 3.22 Typical noise-free heading angle trajectories. (a) Vehicle $k_{j,i}$ with a change of planar orientation at t_1. (b) Vehicle j_i^F with non-abrupt actuator fault occurring at t_f. (c) Vehicle i dynamically coupled to j_i^F.

vehicle $k_{j,i}$ is healthy, from Fig. 3.22 and Definition 3.3. Decision function g is parameterized by the probability of false alarm p_F, and by sample set sizes N_i and $N_{k_{j,i}}$. Function $\mathcal{T}$, which is derived in Sec. 3.4.4, should provide enough information to extract the necessary knowledge on the health status of the vehicle by means of signals $\widehat{\psi}_{k_{j,i}}(kT_s)$ and $\widehat{\psi}_i(kT_s)$. During normal, or healthy, operating conditions, sequences $\mathcal{Y}_{1,N_{k_{j,i}}}$ and $\mathcal{Y}_{1,N_i}$ are correlated. It means that the time trajectory of $\widehat{\psi}_i$ is related to that of $\widehat{\psi}_{k_{j,i}}$ through dynamic coupling entailed by distributed formation control. When a fault occurs, however, there is an informational discrepancy between the two sequences. The comparison of the two sequences is done with the objective of reducing the level of false alarms caused by delayed transients of i after time $t_1 + t_{j/k_{j,i}}$, as shown in Fig. 3.22(c), and by noisy measurements. In Fig. 3.22, $t_{x/y}$ denotes the propagation time naturally occurring when a follower x responds to the motion of its immediate predecessor y being constrained to adhere to the formation control law. Symbol t_1 indicates the time at which the leader starts turning the corner. UAV i detects a non-abrupt fault in $j_i^F \in \mathcal{N}_i$ by analyzing and comparing the impact of a faulty behavior of j_i^F on i. This impact is expected to be detectable owing to dynamic couplings expressed by the edges of G_{FC}. Achieving fault detection by utilizing the propagation through the formation rather than directly detecting the faulty dynamics of the vehicle may result in a slow detection process. This apparent drawback is however acceptable as non-abrupt faults are inherently slow.

3.4.3 *Estimator of heading angle*

As an example of a heading angle estimator, consider once again the ALTAV. Each vehicle i estimates its heading angle by means of the following discrete-time estimator

$$
\begin{aligned}
\gamma^f_{i,k+1} &= e^{-a_\gamma T_s}\gamma^f_{i,k} + (1 - e^{-a_\gamma T_s})\gamma'_i,\\
\phi^f_{i,k+1} &= e^{-a_\phi T_s}\phi^f_{i,k} + (1 - e^{-a_\phi T_s})\phi'_i,\\
X^f_{i,k+1} &= A_{d,f}X^f_{i,k} + B_{d,f}\begin{bmatrix} x'_{i,k+1} \\ y'_{i,k+1} \\ \frac{1}{M}\sum_j F_{ji}\sin(\gamma^f_{i,k}) \\ \frac{1}{M}\sum_j F_{ji}\sin(\phi^f_{i,k}) \end{bmatrix},\\
Y^f_{i,k} &= \begin{bmatrix} Y^1_{i,k} \\ Y^2_{i,k} \end{bmatrix}\\
&= L_f X^f_{i,k},\\
\widehat{\psi}_{i,k} &= \tan^{-1}(Y^2_{i,k}/Y^1_{i,k}),
\end{aligned}
\tag{3.92}
$$

where $1/a_\gamma$ and $1/a_\phi$ are the time constants of the filter applied to γ'_i and ϕ'_i. Subscript k denotes the kth sampling instant with period T_s; that is,

$$\widehat{\psi}_{i,k} = \widehat{\psi}_i(kT_s).$$

Estimator (3.92) is obtained from a zero-order-hold-equivalent discretization, or step-invariant model [139], of the continuous-time observer (A_f, B_f, L_f), where

$$
\begin{aligned}
A_{d,f} &= e^{A_f T_s},\\
B_{d,f} &= A_f^{-1}(1 - e^{A_f T_s})B_f,
\end{aligned}
$$

and

$$L_{d,f} = L_f.$$

Matrices A_f, B_f, and L_f are computed so that the continuous-time observer is robust to slowly time-varying, although bounded, parameters C_x and C_y in Eqs. (3.1). Robustness is achieved by first deriving a polytopic model of the first two equations in Eqs. (3.1), and then solving the linear matrix inequalities of Theorem 2 in Ref. [140] at each vertex of the polytope. In doing so, a robust state estimate is obtained in the minimum variance sense. It is interesting to note that this design is similar to the one presented for the DAFD observer, as applied to system (3.59), although without consideration of exogenous disturbances.

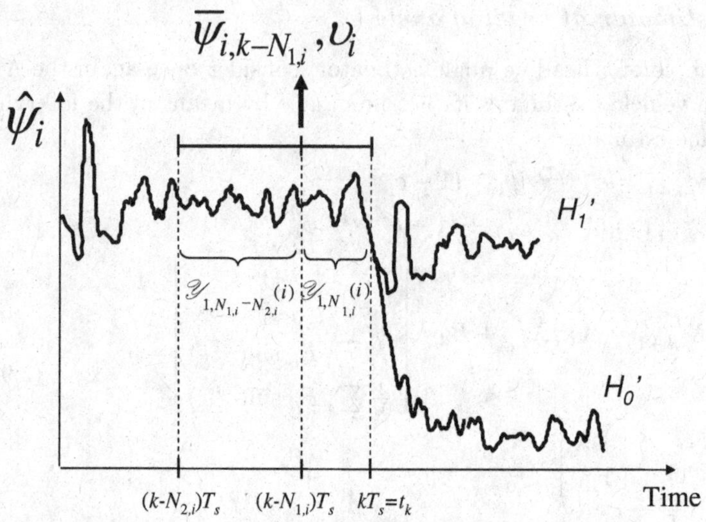

Fig. 3.23   Statistical test performed by DNaFD onboard vehicle i to detect changes in heading angle.

3.4.4   *Statistical test for DNaFD*

Signal-based DNaFD is designed with the hypothesis testing framework described in Refs. [92, 141]. This approach is shown to provide detection of actuator faults of the non-abrupt type defined in Sec. 3.1.3. Since obtaining a closed-form probability distribution of $\widehat{\psi}_i$ from Eqs. (3.92) is arduous, a nonparametric detector is derived from $\widehat{\psi}_i$ [142].

Decision function g of UAV i is based on sequences $\mathcal{Y}_{1,N_i}(i)$ and $\mathcal{Y}_{1,N_{k_{j,i}}}(k_{j,i})$. We first present a statistical test applied to $\mathcal{Y}_{1,N_i}(i)$. The same test applies to $\mathcal{Y}_{1,N_{k_{j,i}}}(k_{j,i})$, which then allows deriving g in Eq. (3.91). As shown in Fig. 3.23, DNaFD onboard vehicle i collects $\widehat{\psi}_i$ over time interval $[(k-N_{2,i})T_s, (k-N_{1,i})T_s]$ to derive empirical information used by g. Then, detection that $\widehat{\psi}_i$ has actually changed over $[(k-N_{1,i})T_s, kT_s]$ yields a result at $t_k = kT_s$. In Fig. 3.23, detection of a change, or behavior marked by fault, occurs with $H_0'(i)$, namely when the empirical frequency is smaller than $p_s(i)$, otherwise a lack of change, or normal behavior, is given by $H_1'(i)$.

Average heading angle $\overline{\psi}_{i,k-N_{1,i}}$ and bound υ_I are obtained empirically from sequence

$$\mathcal{Y}_{1,N_{1,i}-N_{2,i}}(i) = \{\widehat{\psi}_i((k - N_{2,i})T_s), ..., \widehat{\psi}_i((k - N_{1,i})T_s)\}$$

as follows

$$\begin{aligned} &\overline{\psi}_{i,k-N_{1,i}} = \tfrac{1}{N_{2,i}-N_{1,i}} \textstyle\sum_{j=1}^{N_{2,i}-N_{1,i}} \widehat{\psi}_i((k - N_{2,i} + j)T_s), \\ &Q(\mathcal{Y}_{1,N_{1,i}-N_{2,i}}(i) \in [\overline{\psi}_{i,k-N_{1,i}} - v_I, \overline{\psi}_{i,k-N_{1,i}} + v_I]) = \alpha, \end{aligned} \tag{3.93}$$

where $\alpha \in (0, 1)$ is related to the prescribed probability of false alarms, p_F, defined in the sequel. Symbol Q denotes the empirical probability that the elements of $\mathcal{Y}_{1,N_{1,i}-N_{2,i}}(i)$ are in the interval $[\overline{\psi}_{i,k-N_{1,i}} - v_I, \overline{\psi}_{i,k-N_{1,i}} + v_I]$. Recall that the empirical frequency $Q(\{y_1, ..., y_n\} \in A)$, where A is some set, is defined as

$$Q(\{y_1, ..., y_n\} \in A) = \frac{1}{n} \sum_{i=1}^{n} I_A(y_i), \tag{3.94}$$

where I stands for the indicator function

$$I_A(y_i) = \begin{cases} 1 \text{ if } y_i \in A, \\ 0 \text{ otherwise.} \end{cases} \tag{3.95}$$

Since the distribution of $\widehat{\psi}_i$ is not known *a priori*, v_I is difficult to calculate. Yet, one can obtain v_I by means of numerical simulations. As a starting point, we set v_I to 2ρ, where ρ^2 is the sample variance obtained as

$$\rho^2 = \frac{1}{N_{2,i} - N_{1,i} - 1} \sum_{j=1}^{N_{2,i}-N_{1,i}} (\widehat{\psi}_i((k - N_{2,i} + j)T_s) - \overline{\psi}_{i,k-N_{1,i}})^2. \tag{3.96}$$

We thus have a 95% confidence level that $\widehat{\psi}_i$ is element of $[\overline{\psi}_{i,k-N_{1,i}} - v_I, \overline{\psi}_{i,k-N_{1,i}} + v_I]$, provided that $\widehat{\psi}_i$ is Gaussian, which is not necessarily the case however. We then tune v_I by means of extensive simulations such that v_I complies with an acceptable level of false alarms found in the sequel.

Let

$$p_s(i) = \frac{n_{s,i}}{N_{1,i}(i)}.$$

Furthermore, let $H_1'(i)$ and $H_0'(i)$ stand for time-constant and time-varying heading angle trajectories, respectively, of vehicle i. The statistical test to decide whether

$$\mathcal{Y}_{1,N_{1,i}}(i) = \{\widehat{\psi}_i((k - N_{1,i})T_s), ..., \widehat{\psi}_i(kT_s)\}$$

is in the interval $[\overline{\psi}_{i,k-N_{1,i}} - \upsilon_I, \overline{\psi}_{i,k-N_{1,i}} + \upsilon_I]$ with a probability of false alarms $P_F \leq p_F$ is given by

$$\widehat{P}(\mathcal{Y}_{1,N_{1,i}}(i) \in [\overline{\psi}_{i,k-N_{1,i}} - \upsilon_I, \overline{\psi}_{i,k-N_{1,i}} + \upsilon_I]) \underset{H_0'(i)}{\overset{H_1'(i)}{\gtrless}} p_s(i), \tag{3.97}$$

where $\widehat{P}$ denotes the empirical frequency defined in Eq. (3.94) and the probability of false alarms is related to p_s as

$$\begin{aligned} P_F &= E\left[\textstyle\sum_{u=1}^{N_{1,i}} I_{[\overline{\psi}_{i,k-N_{1,i}} - \upsilon_I, \overline{\psi}_{i,k-N_{1,i}} + \upsilon_I]}(\widehat{\psi}_i((k - N_{1,i} - u + 1)T_s)) > n_{s,i}\right] \\ &= \textstyle\sum_{i=n_{s,i}}^{N_{1,i}} \binom{N_{1,i}}{i} \alpha^i (1-\alpha)^{N-i}. \end{aligned} \tag{3.98}$$

With reference to Fig. 3.23, the probability of false alarms $\widehat{P}$ in (3.97) is obtained at time t_k. In Eq. (3.98), $E[\cdot]$ denotes the expected value of its argument. Inequality$P_F \leq p_F$ is thus obtained provided $n_{s,i}$ $(= p_s(i)N_{1,i})$ satisfies

$$\sum_{i=n_{s,i}}^{N_{1,i}} \binom{N_{1,i}}{i} \alpha^i (1-\alpha)^{N-i} \leq p_F, \tag{3.99}$$

where

$$\binom{N_{1,i}}{i}$$

stands for the binomial coefficient.

The procedure given in (3.93)-(3.99), with

$$p_s(k_{j,i}) = \frac{n_{s,k_{j,i}}}{N_{1,k_{j,i}}(k_{j,i})},$$

is applied to $\mathcal{Y}_{1,N_{1,k_{j,i}}-N_{2,k_{j,i}}}(k_{j,i})$ and $\mathcal{Y}_{1,N_{1,i}}(i)$ over $[(k - N_{2,i})T_s, (k - N_{1,i})T_s]$ and $[(k - N_{1,i})T_s, kT_s]$, respectively, which leads to the selection of $H_1'(k_{j,i})$ or $H_0'(k_{j,i})$. We are now ready to define $\mathcal{T}$, which is instrumental to the decision function g in Eq. (3.91), as follows

$$\mathcal{T}(t_k, \mathcal{Y}_i) = \begin{cases} 1 & \text{if } \begin{cases} H_0'(i) \text{ at } t = t_k \text{ and } H_1'(k_{j,i}) \text{ for all } t \in [t_k - t_{dc}, t_k], \\ \text{or} \\ H_1'(i) \text{ at } t = t_k \text{ and } H_0'(k_{j,i}) \text{ for all } t \in [t_k - t_{dc}, t_k], \end{cases} \\ 0 & \text{if } \begin{cases} H_1'(i) \text{ at } t = t_k \text{ and } H_1'(k_{j,i}) \text{ for all } t \in [t_k - t_{dc}, t_k], \\ \text{or} \\ H_0'(i) \text{ at } t = t_k \text{ and } H_0'(k_{j,i}) \text{ for all } t \in [t_k - t_{dc}, t_k]. \end{cases} \end{cases} \tag{3.100}$$

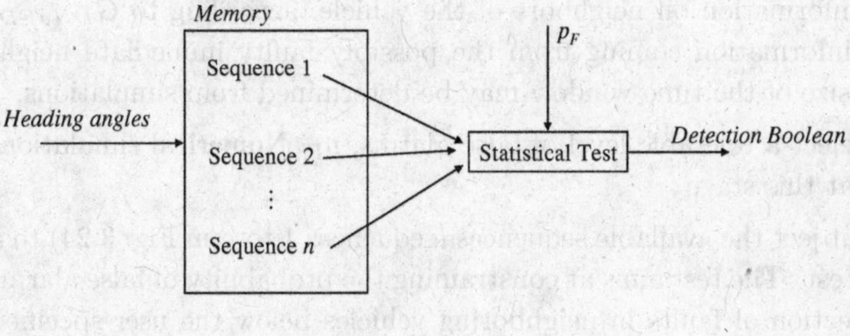

Fig. 3.24 Schematics of DNaFD.

In Eq. (3.100), t_{dc} stands for the propagation time of heading angle transients between $k_{j,i}$ and i. Vehicle i is dynamically coupled to $k_{j,i}$ through j_i. Thus, t_{dc} satisfies $t_{dc} \geq t_{i/j_i} + t_{j_i/k_{j,i}}$, where t_{i/j_i} and $t_{j_i/k_{j,i}}$ are depicted in Fig. 3.23. The decision function verifies whether dynamically coupled vehicles have similar behavior up to the propagation time t_{dc}.

3.4.5 *DNaFD design algorithm*

The steps in the design of the DNaFD function are given as follows. The structure of a DNaFD scheme is shown in Fig. 3.24. Heading angle information obtained over a certain time window is processed and stored in memory as a set of sequences. There is one sequence of heading angles corresponding to a vehicle per time window. A statistical test is performed on those sequences to determine whether a fault has occurred on each of the immediate predecessor vehicles.

Step 1. For a given number of vehicles, a mission of interest, and a formation control law, determine if the information flow graph ensures a minimum number of signals for DNaFD operation. In other words, verify that the nodes on which DNaFD is to be implemented have indegree greater than or equal to two, and that Definition 3.3. is satisfied. When these conditions are violated, DNaFD is inoperative.

Step 2. With the available onboard sensors and the processing capability, obtain measurements of the attitude of the vehicle for a given time window, and design an estimator of the type given in Eqs. (3.92). Store in memory the heading angle information as a sequence of length N_i.

Step 3. Over the time window of interest, design a system to collect heading angle information on neighbors of the vehicle, according to G_{DNaFD}, excluding information coming from the possibly faulty immediate neighbors. The size of the time window may be determined from simulations.

Step 4. Select a tolerable level of false alarms, p_F. Numerical simulations are useful at this stage.

Step 5. Subject the available sequences (sequences 1 to n in Fig. 3.24) to a statistical test. The test aims at constraining the probability of false alarms on the detection of faults in neighboring vehicles below the user-specified level p_F. Implement the decision function g, given in Eq. (3.91). From Eq. (3.100), the detection boolean is set to unity when a faulty behavior is detected beyond the false-alarm level, it is zero otherwise.

3.5 UAV Command Adaptation

Once the DFD detects an anomalous behavior using combined DAFD and DNaFD systems, an adaptation process takes place, as indicated in the lower part of Fig. 3.16 for DAFD. Here, we present a command adaptation scheme obtained by means of straightforward decision rules. Once a fault on an immediate predecessor vehicle is detected by means of the DAFD and DNaFD functions, a flag is raised. The detection of a fault is a boolean entering the flight control system of the vehicle and tells the vehicle that it should no longer follow its faulty predecessor. Instead, the UAV is told to track the nearest healthy UAVs within range and to adapt its control system to achieve the required relative positions and speeds with respect to those healthy predecessor vehicles. Clearly, however, there will be a limit to the extent of team recovery if several UAVs are at fault.

The principle of command adaptation is shown in Figs. 3.25 and 3.26. Figure 3.25 represents a number of follower UAVs in a formation at a given time instant. UAV n has a certain sensor range represented by a circle. UAVs $n-1$, $n-2$ and $n-3$ are within range of UAV n. The formation control law requires UAV n to follow UAV $n-1$. The arrow from UAV $n-1$ to UAV n indicates that the latter has access to information on vehicle $n-1$ used to guide its own flight under nominal operating conditions. However, during mission, UAV $n-1$ experiences a fault. Systems DAFD and DNaFD, onboard UAV n, detect the problem after some time. Then UAV n uses its onboard sensors to search for the nearest healthy UAV within range, which is UAV $n-2$ in Fig. 3.25, and tries to maintain the required relative separation from it, disregarding data from UAV $n-1$. The formation

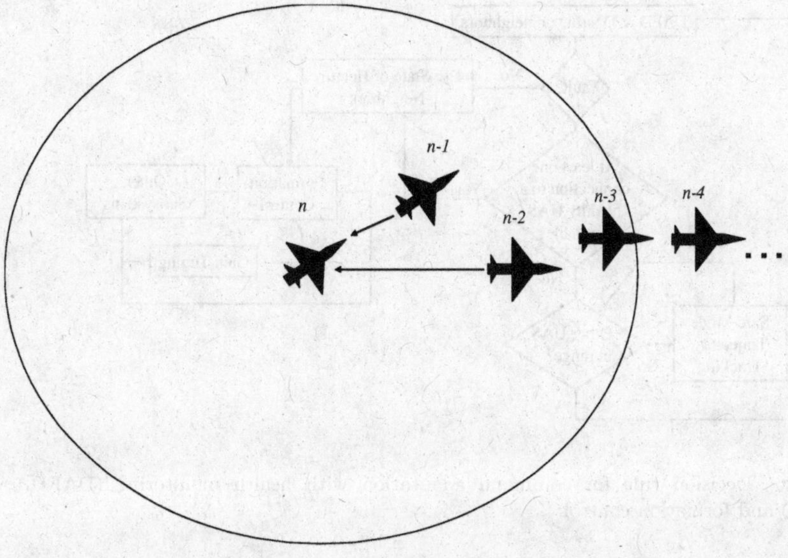

Fig. 3.25 Command adaptation for UAV n.

control system onboard UAV n relies on the information coming from UAV $n-2$ to continue its flight, while the relative motion of the rest of the fleet is unaffected by this local adaptation. At a higher level of the cooperative control system hierarchy, the loss of one or more UAVs may necessitate adaptation of the team. This will be discussed in more detail in Chapter 4.

Figure 3.26 presents a command adaptation algorithm. Its outcome is the selection of the state trajectories to be tracked by the formation control law onboard the UAS and the parameters of the formation control law, or simply the controller gains. The algorithm can be explained as follows. On any given UAV i, the DAFD and DNaFD functions run at a certain update rate. They carry out monitoring on the neighboring vehicles (vehicles within range of the sensors) and on those for which the network provides state information when available. Every neighboring vehicle of i is thus monitored. When no fault is detected on any of those vehicles, the formation control law embedded onboard UAV i simply uses the relevant formation control information on the immediate neighbors of i, namely those vehicles directly connected to i. However, when a fault is detected, the algorithm checks to see if there is currently at least one connection to a healthy neighboring vehicle. If so, the formation control law uses

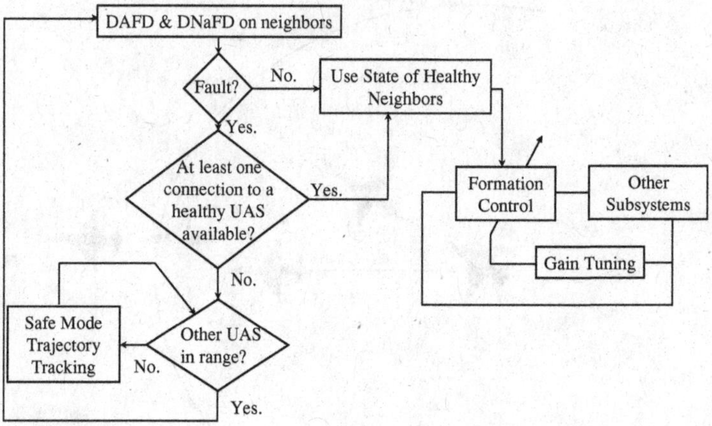

Fig. 3.26 Decision rule for command adaptation with health monitoring (DAFD, DNaFD) and formation control.

the state information coming from the healthy immediate neighbors, and not that from vehicles marked as faulty. Controller gains may have to be tuned at this stage and, in addition, if the fleet can no longer maintain the formation geometry obtained in nominal operating conditions, a higher-level decision may have to be made to reconfigure it. At this point, we concentrate on the simple algorithm in Fig. 3.26. If there is no connection to a healthy UAS available to vehicle i, due to multiple faults for instance, the system must determine if there is a nearby UAS within range, whether it is an immediate neighbor of vehicle i or not. If there is at least one healthy vehicle, the DAFD and DNaFD detectors are applied to it, provided sufficient information is available, and the process continues. If there is no UAS within range, the vehicle is commanded to go into what is known as safe mode trajectory tracking.

As it is no longer running the nominal formation control law, the vehicle may go astray with respect to the rest of the formation. The UAS may be embedded with the functionalities of safe return to base, search, loiter, or hover, for example, provided the control authority allows for such maneuvers. Meanwhile, the UAS continues looking for teammates using its available onboard sensors. Again, if a UAS finds a healthy neighboring vehicle to follow, the formation controller gains may have to be tuned. A lookup table may be stored in memory to speed up online tuning if the number of potential cases is relatively small.

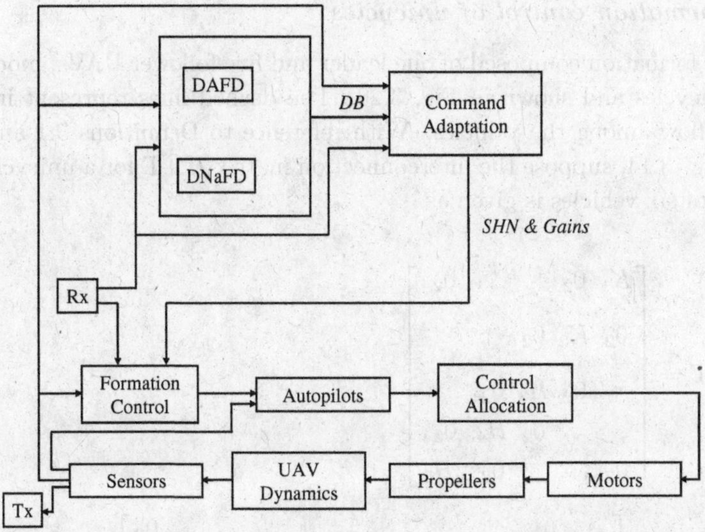

Fig. 3.27 Block diagram of DFD-CA and closed-loop system for follower UAV.

A block diagram of the DFD-CA and formation control systems is shown in Fig. 3.27. Blocks represent DAFD, DNaFD, command adaptation, UAV dynamics, low-level control, actuators, sensors, Tx/Rx and formation control. DB stands for detection boolean, SHN for the state of healthy neighbors, and Gains signifies formation controller gains. The figure presents a complete picture of the control system, incorporating loops already described in Figs. 3.4, 3.6, 3.7, and 3.26.

3.6 Simulations and Experiments

Numerical simulations and experiments illustrate the effectiveness of the formation control, health monitoring and command adaptation functions. Taking a similar step-by-step approach as in Sec. 3.2, we begin by assuming that each UAV behaves as a unicycle. Then, we present more realistic nonlinear dynamic models of UAVs and conditions of operation, where some of the parameters of the ALTAV model have been obtained by means of flight experiments.

3.6.1 Formation control of unicycles

Consider a formation composed of one leader and five follower UAVs, modeled as unicycles and shown in Fig. 3.28. The dashed lines represent information flow among the vehicles. With reference to Definitions 3.1 and 3.2, and Fig. 3.14, suppose the interconnection matrix $H+\Gamma$ for a unicycle formation of six vehicles is given as

$$H+\Gamma = \begin{bmatrix} K_1 & 0_2 & & \dots & 0_2 \\ 0_2 & K_2 & 0_2 & & \vdots \\ \vdots & 0_2 & H_3 & 0_2 & \\ & & 0_2 & H_4 & 0_2 \\ 0_2 & \dots & & 0_2 & 2H_5 \end{bmatrix} + \begin{bmatrix} 0_2 & & & \dots & 0_2 \\ 0_2 & 0_2 & & & \vdots \\ -(H_3+I_2) & 0_2 & 0_2 & & \\ -(H_4+I_2) & 0_2 & \dots & 0_2 & \\ 0_2 & -(H_5+\frac{I_2}{2}) & 0_2 & -(H_5+\frac{I_2}{2}) & 0_2 \end{bmatrix}. \tag{3.101}$$

Nodes 1 and 2 are connected to the leader. Following Definition 3.2, controller gains are denoted with the symbol K_i for $i=1,2$. The remaining followers are not connected to the leader and hence bear the symbol H_i, $i=3,4,5$. To understand the content of the interconnection matrix, the reader is referred to Fig. 3.28 which shows the leader, the five followers, and the information flow for formation control.

Control law of the leader vehicle is given as follows. Parameters α and β of the feedback linearizing loop, given in Eqs. (3.17), are set to 0.2 and 0.55, respectively. This feedback linearizing control law is used for the follower unicycles as well. Gains k_p and k_d in Eqs. (3.4) are set to 1.4286 and 1.4, respectively.

Let us focus on the control law for the follower unicycles. Using the MATLAB® LMI toolbox [108], the distributed controller gain SPR matrices $H_i, K_i \in \mathbb{R}^{2\times 2}$ are calculated from LMI (3.45), which serve as constraints in the minimization of

$$L^* = \arg \min_{L\in\{H_i,K_i\}} \|L\|. \tag{3.102}$$

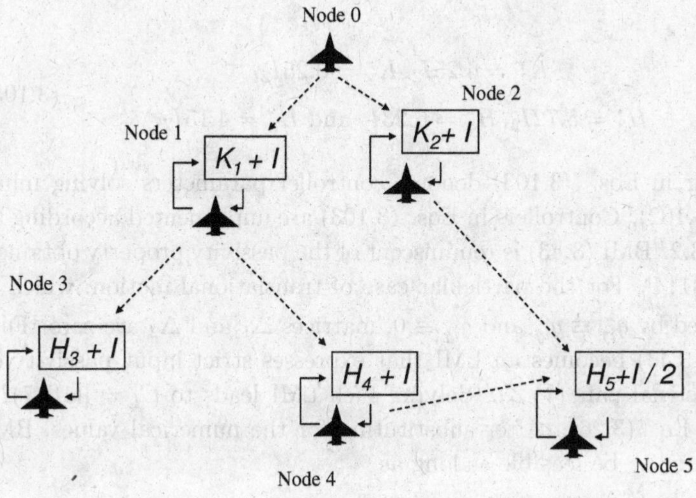

Fig. 3.28 Formation control example.

Another LMI is added to bound the norm of the input of the linearized dynamics (3.29) below $u_{\max}$; that is,

$$||YC\widetilde{Z}||^2 \leq u_{\max},$$

where (1) $C = diag(C_1, ..., C_n)$ is computed by solving BMI (3.43), and (2) matrix Y is obtained by replacing the $(3,3)$, $(4,4)$ and $(5,5)$ entries of $H + \Gamma$ by $(H_3 + I_2)$, $(H_4 + I_2)$ and $(2H_5 + I_2)$, respectively. Following Ref. [107], the input constraint is satisfied if

$$\begin{bmatrix} u_{\max}^2 I & YC \\ C^T Y^T & W^{-1} \end{bmatrix} \geq 0,$$

where matrix $W > 0$ is such that

$$\widetilde{Z}^T W^{-1} \widetilde{Z} \leq 1$$

for all $\widetilde{Z}$ in the contraction region. Selecting

$$u_{\max} = 20\sqrt{10}$$

and

$$W = diag(10, ..., 10)$$

yields

$$\begin{aligned} K_1^* = 6.25 I_2, K_2^* = 6.25 I_2, \\ H_3^* = 5.72 I_2, H_4^* = 6.23 I_2 \text{ and } H_5^* = 4.15 I_2. \end{aligned} \tag{3.103}$$

The star in Eqs. (3.103) denotes controller parameters solving minimization (3.102). Controllers in Eqs. (3.103) are implemented according to Definition 3.2. BMI (3.43) is reminiscent of the passivity property obtained for node i [114]. For the particular case of translational motion, which is characterized by $q_{ij}^* \equiv q_{ij}^o$ and $\dot{q}_{ij}^* \equiv 0$, matrices Δ_1 and Δ_2 are zero. Furthermore, (3.43) becomes an LMI that expresses strict input passivity of node i for signal pair $(V, \widetilde{Z})$. Solving such LMI leads to $C_i = [(10/7)I_2, (7/5)I_2]$ in Eq. (3.26). After substituting for the numerical values, BMI (3.43) is found to be feasible as long as

$$\frac{\rho_{ij}^*}{\|q_i\|^2} < 0.95.$$

The latter is satisfied if $v_m > 1.1\rho_{ij}^*$ from

$$\left|\dot{x}_i\right| \cdot \left\|\dot{q}_i\right\|^{-2} < \frac{1}{v_m}, \quad \left|\dot{y}_i\right| \cdot \left\|\dot{q}_i\right\|^{-2} < \frac{1}{v_m}. \tag{3.104}$$

The right-hand side bounds in inequalities (3.104) are more conservative conditions than those expressed in inequalities (3.40). The simulations rely on the controllers of Eq. (3.103). The six-unicycle formation shown in Fig. 3.28 has for objective to track a sinusoidal trajectory given as

$$q^*(t) = \begin{bmatrix} 14.1t \\ 180\sin(\pi t/60) \end{bmatrix}$$

over $[0 \text{ s}, 120 \text{ s}]$. Furthermore, the formation is required to maintain all relative positions, ρ_{ij}^*, to a value of 5 m during interval $[0, 60)$ seconds and a value of 10 m over $[60, 120)$ seconds while the LOS angle, λ_{ij}^*, is set to a fixed value of $\pi/4$ radians. The change in commanded geometries acts as a disturbance, according to Assumption 3.4. The step command in ρ_{ij}^* at a time of 60 seconds is however smoothed out by passing it through a second-order filter with parameters $\omega_n = 10$ rad/s, and $\xi = 0.7$. The initial operating conditions are $v_i(0) = 10$ m/s, $\theta_i(0) = 0$ radian, and $q_0(0) = [2, -2]^T$ m. The initial positions are computed with $\rho_{ij}^* = 5$ m and

$\lambda_{ij}^* = \pi/4$ radians, and $\theta_i = 0$ for $i = 0, ..., 5$, which gives

$$\begin{aligned}
x_1(0) &= 2 - 5\cos(\pi/4),\\
x_2(0) &= 2 - 5\cos(\pi/4),\\
x_3(0) &= 2 - 10\cos(\pi/4),\\
x_4(0) &= 2 - 10\cos(\pi/4),\\
x_5(0) &= 2 - 10\cos(\pi/4),\\
y_1(0) &= 2 + 5\cos(\pi/4),\\
y_2(0) &= 2 - 5\cos(\pi/4),\\
y_3(0) &= 2 + 10\cos(\pi/4),\\
y_4(0) &= 0,\\
y_5(0) &= 2 - 10\cos(\pi/4).
\end{aligned}$$

To provide a realistic simulation, commands ρ_{ij}^* are transmitted with a delay of one second from one node to another starting from the leader. Yet, if geometry requirements are specified prior to mission, and embedded onboard each vehicle, the delay can be assumed to be zero. Figure 3.29 shows the evolution of the formation in terms of vehicle separation and $x-y$ plane motion. The unicycles experience noise-free measurements. For the sake of clarity, only the trajectory of the leader and of nodes 3 and 5 are shown in part (a). Time instants t_1, t_2 and t_3 correspond to 30 seconds, 75 seconds and 110 seconds, respectively. In part (b) of Fig. 3.29, vehicle separations and headings are seen to be relatively close to the commanded values. In the figure, the arrow represents the velocity vector of each vehicle. Quantitatively, for each node, tracking errors are defined as the differences between required and actual positions on the x and y axes.

Figure 3.30(a) shows the worst and the best time histories of relative errors in positions. The commanded change in formation geometry goes through a low-pass filter and occurs at a time of 60 seconds. The tracking errors shown in Fig. 3.30(a) confirm, along with Fig. 3.29, that the formation geometry is preserved and that each vehicle follows a sinusoidal trajectory with a ratio $\max \|\tilde{q}_i\| / \rho_{ij}^*$ below 10%. Despite transients occurring right after the geometry command change at a time of 60 seconds, the formation remains stable. The ratio reaches its peak value at the extremum of the sinusoidal trajectory; that is, as the desired normal acceleration reaches its peak. This phenomenon corroborates the convergence analysis since the trajectory q_i of each vehicle approaches a ball centered at a desired trajectory q_i^* and having a radius that increases with $d^2\theta_0/dt^2$.

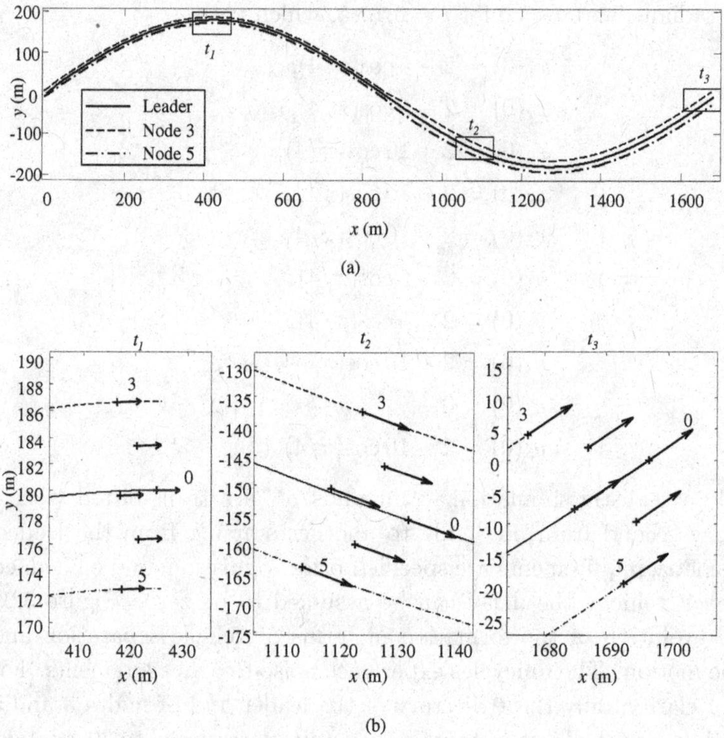

Fig. 3.29 (a) Motion of leader (node 0) and follower nodes 3 and 5 on the $x-y$ plane. (b) Formation geometries at three time instants.

After a peak of 38 m/s, which is due to nonzero initial errors, the control inputs of Fig. 3.30(b) remain within $[0, 20]$ m/s and $[-1, 1]$ radian, for all values of ρ_{ij}^*. If the demands in geometry are increased to $\rho_{ij}^*(0) = 20$ m and $\rho_{ij}^*(60) = 40$ m, the formation remains stable, as shown in Fig. 3.31.

For the situations depicted in Figs. 3.30 and 3.31, the geometry commands enter a second-order filter with parameters $\omega_n = 2$ rad/s and $\xi = 1$. However, if geometry commands of $\rho_{ij}^*(0) = 20$ m and $\rho_{ij}^*(60) = 40$ m are passed through a large-bandwidth filter, namely a second-order filter with $\omega_n = 5$ rad/s, $\xi = 0.7$, the formation behaves in an unacceptable manner, as shown in Fig. 3.32. Briefly, a filter characterized by a large bandwidth leads to erroneous responses whereas a filter with a smaller bandwidth leads to stable formation flight. Correctly filtering geometry commands is thus

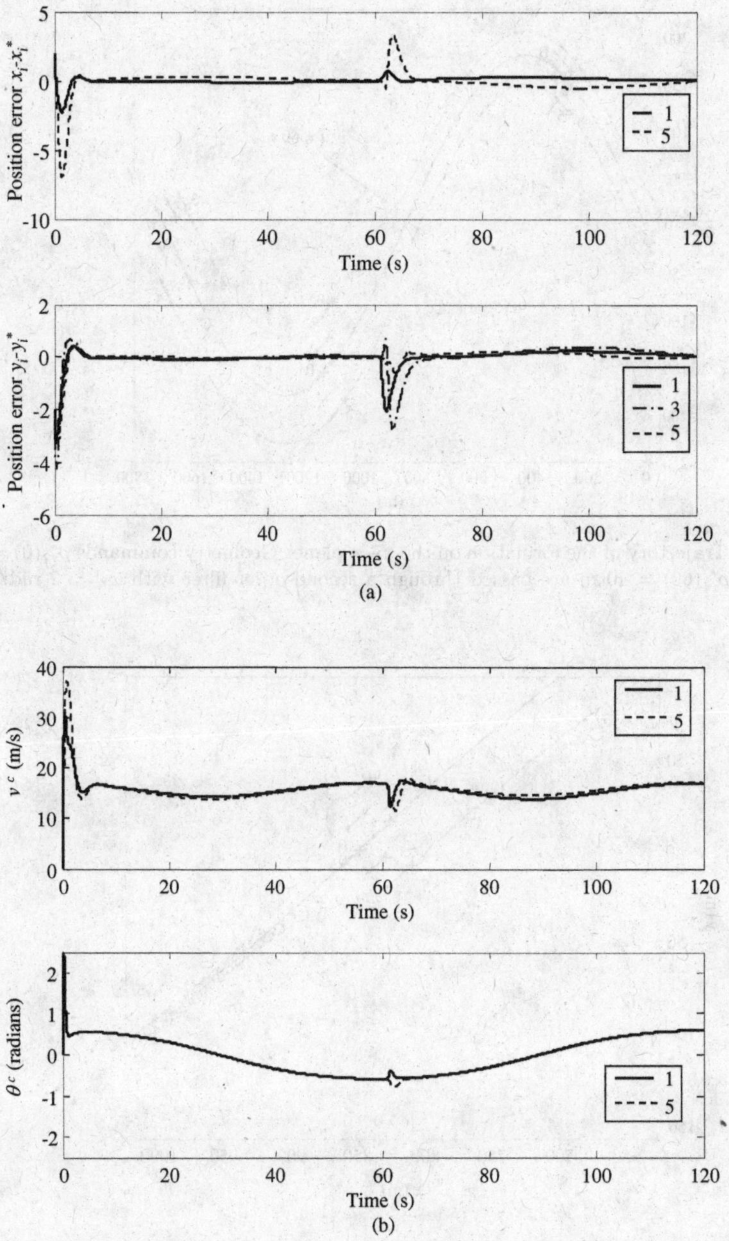

Fig. 3.30 (a) Worst and best tracking errors. Command change at 60 seconds goes through a second-order filter with parameters $\omega_n = 2$ rad/s and $\xi = 1$. (b) Control signals (3.25) of nodes 1 and 5.

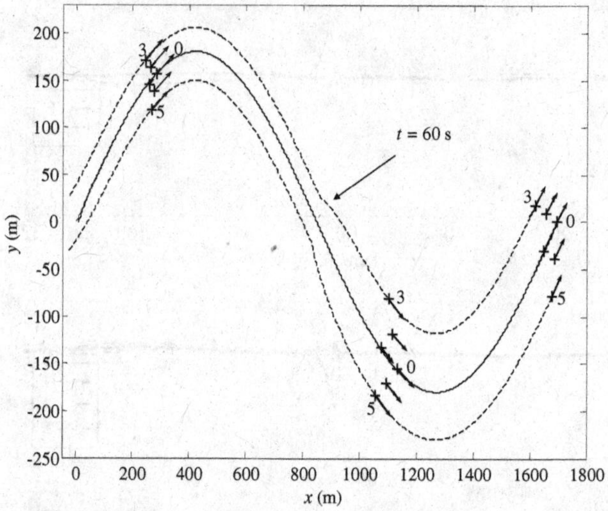

Fig. 3.31 Trajectory of the formation on the $x-y$ plane. Geometry commands $\rho^*_{ij}(0) =$ 20 m and $\rho^*_{ij}(60) = 40$ m are passed through a second-order filter with $\omega_n = 2$ rad/s and $\xi = 1$.

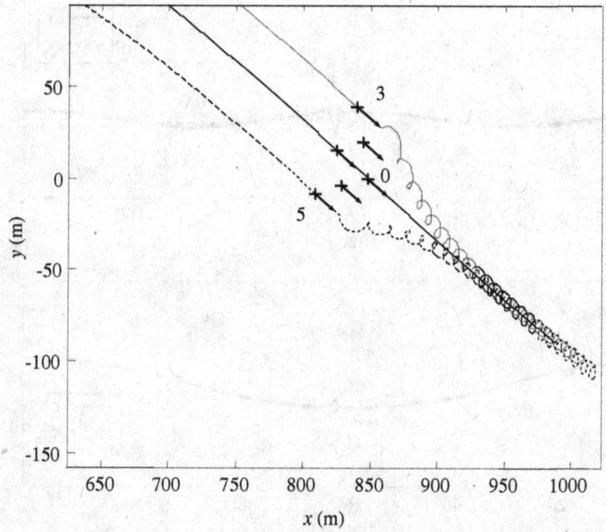

Fig. 3.32 Trajectory of the formation on the $x-y$ plane. Geometry commands $\rho^*_{ij}(0) =$ 20 m and $\rho^*_{ij}(60) = 40$ m are passed through a large-bandwidth second-order filter with $\omega_n = 5$ rad/s and $\xi = 0.7$.

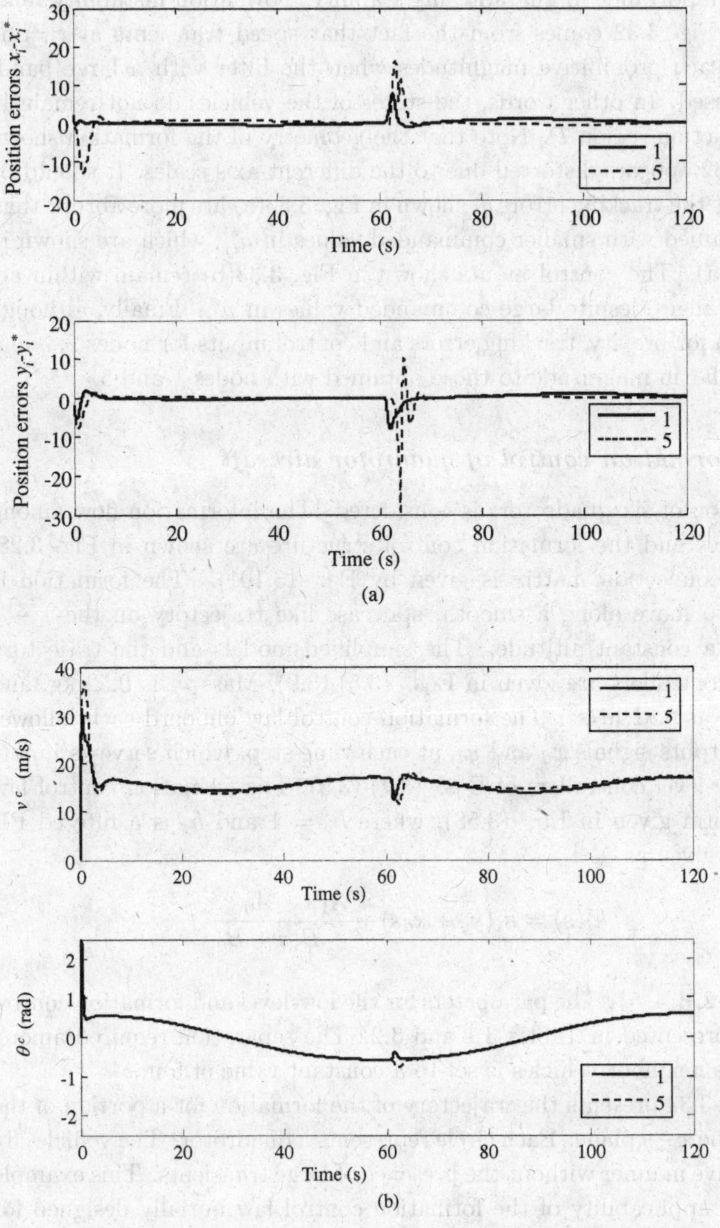

Fig. 3.33 (a) Tracking errors for nodes 1 and 5, and (b) control inputs for commands $\rho_{ij}^*(0) = 20$ m and $\rho_{ij}^*(60) = 40$ m, which are passed through a second-order filter with parameters $\omega_n = 2$ rad/s and $\xi = 1$.

of prime importance in guaranteeing stability. Formation instability illustrated in Fig. 3.32 comes from the fact that speed transients at $t = 60$ seconds reach prohibitive magnitudes when the filter with a large bandwidth is used. In other words, the states of the vehicles do not remain in the contraction region $\widetilde{\mathcal{D}}$. Note that the geometry of the formation shown in Fig. 3.32 appears distorted due to the different axis scales. It should be noted that the tracking errors $\widetilde{q}_i$ shown in Fig. 3.33(a) are more abrupt than those obtained with smaller commanded values in ρ^*_{ij}, which are shown in Fig. 3.30(a). The control inputs shown in Fig. 3.33(b) remain within acceptable ranges despite large commanded values in ρ^*_{ij}. Finally, although not shown for brevity, tracking errors and control inputs for nodes 2, 3 and 4 are similar in magnitude to those obtained with nodes 1 and 5.

3.6.2 *Formation control of quadrotor aircraft*

A formation of six quadrotors is considered. The information flow among the vehicles and the formation control structure are shown in Fig. 3.28. The interconnection matrix is given by Eq. (3.101). The formation is required to move along a smooth, staircase like trajectory on the $x - y$ plane, at a constant altitude. The simplified models and the trajectory tracking controllers are given in Eqs. (3.5)-(3.9). Mass m is 0.52 kg, and g is taken as 9.81 m/s^2. The formation control law onboard each follower vehicle outputs signals x_d and y_d, at each time step, which serve as inputs to the low-level controllers of Eqs. (3.7)-(3.9). The formation control law has the form given in Eq. (3.58), where $k_i = 1$ and h_i is a filtered PD controller given as

$$h_i(s) = \kappa_i(k_p + k_d s)\frac{A_1 s + A_0}{s^2 + B_1 s + B_0}$$

for $i \in \{1, 2, 3, 4, 5\}$. The parameters for the low-level and formation control laws are presented in Tables 3.1 and 3.2. The separation required among immediate neighbor vehicles is set to a constant value of 5 m.

Figure 3.34 presents the trajectory of the formation for a portion of the flight on the $x-y$ plane. Each circle represents a quadrotor. The vehicles fly in a cohesive manner without the presence of large transients. This example shows the applicability of the formation control law initially designed for simple unicycle models to nonlinear dynamic systems of various types. The structure of the formation control law remains unchanged, although tuning of the controller parameters is required when going from a unicycle model to a more complex model.

Table 3.1 Parameters for low-level control.

$k_p = 2$	$k_d = 2$	$k_i^c = 0$ (leader)
$k_i^c = 0.002$ (followers)	$a_{z1} = 0.1$	$a_{z2} = 0.01$
$\lvert\sigma_{\phi 1}\rvert \leq 2$	$\lvert\sigma_{\phi 2}\rvert \leq 1$	$\lvert\sigma_{\phi 3}\rvert \leq 0.2$
$\lvert\sigma_{\phi 4}\rvert \leq 0.1$	$\lvert\sigma_{\theta 1}\rvert \leq 2$	$\lvert\sigma_{\theta 2}\rvert \leq 1$
$\lvert\sigma_{\theta 3}\rvert \leq 0.2$	$\lvert\sigma_{\theta 4}\rvert \leq 0.1$	

Table 3.2 Parameters for formation control law.

$k_p = 2$	$k_d = 2$	$K_1 = \kappa_1 \cdot I_2$
$K_2 = \kappa_2 \cdot I_2$	$H_3 = \kappa_3 \cdot I_2$	$H_4 = \kappa_4 \cdot I_2$
$H_5 = \kappa_5 \cdot I_2$	$\kappa_1 = 271.09$	$\kappa_2 = 125.42$
$\kappa_3 = 74.39$	$\kappa_4 = 119.35$	$\kappa_5 = 41.96$
$A_1 = 2 \times 10^3$	$A_0 = 4 \times 10^6$	$B_1 = 4 \times 10^3$
$B_0 = 4 \times 10^6$		

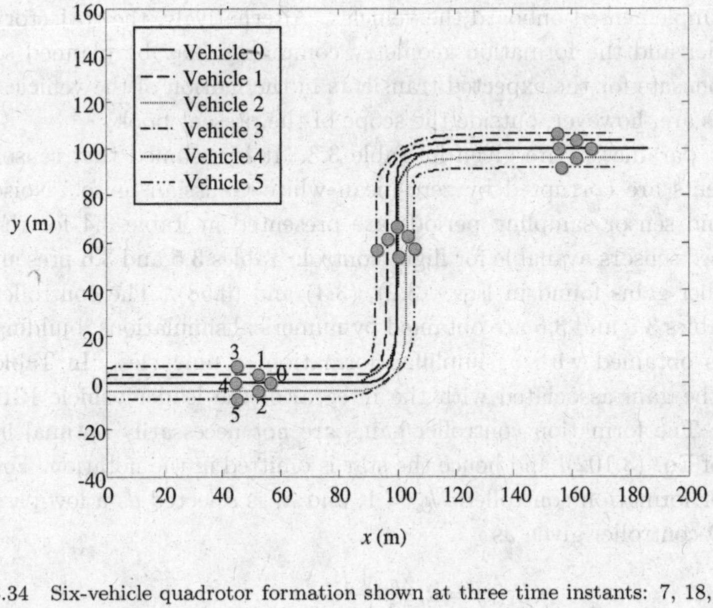

Fig. 3.34 Six-vehicle quadrotor formation shown at three time instants: 7, 18, and 28 seconds.

3.6.3 *DAFD and formation control of ALTAVs*

Consider a formation of nine ALTAVs. Node 1 is the leader, whereas nodes 2 to 9 constitute the followers. The formation is required to fly in a string-type geometry with regular spacing between successive vehicles. Neighboring set N_i is defined, for all $i \in \{2, ..., 9\}$, as $N_i = \{i - 1\}$. Therefore, the information required for formation flight flows from node 1 to node 2, and then from node 2 to node 3, and so on. The interconnection matrix $H + \Gamma$ for this ALTAV formation is given as

$$H + \Gamma = diag\left(\begin{bmatrix} \kappa_2 I_2 & 0_2 \\ -(\kappa_3 + 1) I_2 & \kappa_3 I_2 \end{bmatrix}, ..., \begin{bmatrix} \kappa_8 I_2 & 0_2 \\ -(\kappa_9 + 1) I_2 & \kappa_9 I_2 \end{bmatrix}\right).$$

In the expression for $H + \Gamma$, 0 denotes a matrix of dimension 2×2 filled with zeros. As depicted in Fig. 3.35, the formation moves along a square-type trajectory on the $x - y$ plane. The square has side length of 100 m and the vehicles fly at a speed of around 1 m/s. The triangles indicate the positions of the leader at five different time instants. Followers are required to stabilize around commands $\rho^* = 1$ m and $\lambda^* = 0$ radian. However, follower trajectories are not constrained beyond the stabilization requirement. The latter implies that an obstacle/collision avoidance scheme has to be implemented onboard the vehicles. Alternatively, the trajectory of the leader and the formation geometry commands may be planned so as to compensate for the expected transients in the motion of the vehicles. Such topics are, however, outside the scope of the present book.

ALTAV parameters are given in Table 3.3. It is assumed that sensor measurements are corrupted by zero-mean white Gaussian noise. Noise variance and sensor sampling periods are presented in Table 3.4 for the four onboard sensors available for flight control. Tables 3.5 and 3.6 present the controller gains found in Eqs. (3.2), (3.4) and (3.58). The controller gains of Tables 3.5 and 3.6 are obtained by numerical simulations, building upon gains obtained with a simplified formation of unicycles. In Table 3.5, k_i is the gain associated with the integral of the leader vehicle PID controller. The formation controller gains are not necessarily optimal in the sense of Eq. (3.102), and hence the star is omitted in the notation. For the follower formation controllers, $k_i = 1$, and h_i is selected as a low-pass filtered PD controller given as

$$h_i(s) = \kappa_i \frac{k_{p_i} + k_{s_i} s}{1 + \tau_i s}.$$

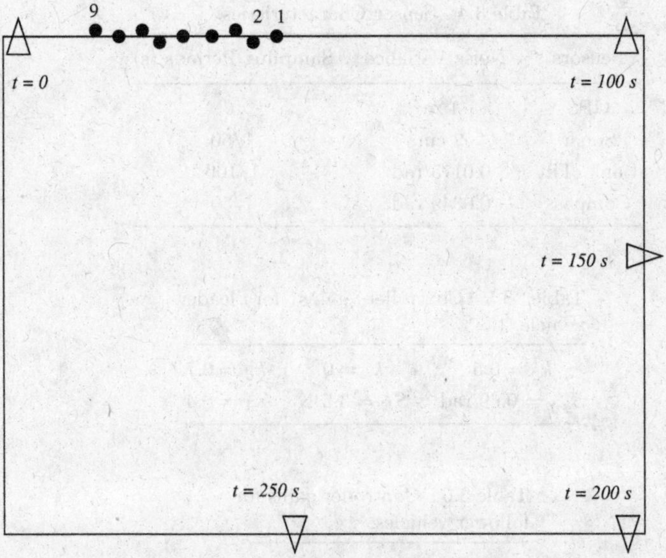

Fig. 3.35 Formation of 9 vehicles along square-type trajectory.

Table 3.3 ALTAV parameters.

$F_B = 13\text{N}$	$M = 1.618\text{kg}$
$J_t = 0.995\text{kg.m}^2$	$J = 1.005\text{kg.m}^2$
$l_1 = 0.941\text{m}$	$l_2 = 0.941\text{m}$
$l_3 = 0.941\text{m}$	$l_4 = 0.941\text{m}$
$l_b = 0.16\text{m}$	$g = 9.8\text{m/s}^2$
$C_x = 0.95$	$C_y = 0.95$
$C_z = 0.95$	$C_t = 0.5$
$C_g = 0.5$	$C_p = 0.5$

The ALTAV formation operates under degraded conditions. Indeed, a hard-over fault of a single actuator occurs at a time of fault t_f of 150 seconds onboard vehicles 2 and 6 concurrently. The HOF actuator fault is described in Sec. 3.1.3 and illustrated in Fig. 3.9. We assume that the FDD-FTC system is either absent or inoperative for the individual vehicle. For both vehicles 2 and 6, actuators saturate at 3.5 N; that is, one has from

Table 3.4 Sensor Characteristics.

Sensors	Noise Variance	Sampling Periods (s)
GPS	1 m^2	1
Sonar	2 cm^2	1/50
Bank, Tilt	0.0175 rad^2	1/100
Compass	0.0349 rad^2	1/50

Table 3.5 Controller gains for leader vehicle.

$k_p = 0.5$	$k_i = 0$	$k_d = 0.7$
$S_{\varphi\gamma} = 0.69$ rad	$S_F = 3.5$ N	$\kappa_1 = 0.4$

Table 3.6 Controller gains for follower vehicles.

$p_z = 0.6$	$i_z = 0.2$
$d_z = 5$	$p_{\varphi\gamma} = 1$
$i_{\varphi\gamma} = 0$	$d_{\varphi\gamma} = 3$
$p_\theta = 0.125$	$i_\theta = 0.1$
$d_\theta = 1.5$	$k_{p_i} = 0.5$
$k_{s_i} = 0.7$	$S_{\varphi\gamma} = 0.69$ rad
$S_F = 3.5$ N	$\kappa_2 = 0.3$
$\kappa_3 = 0.3$	$\kappa_4 = 0.25$
$\kappa_5 = 0.23$	$\kappa_6 = 0.23$
$\kappa_7 = 0.21$	$\kappa_8 = 0.21$
$\kappa_9 = 0.2$	$\tau_i = 0.05$ s

Eqs. (3.10) and (3.11)

$$\delta_i = 1_{t-t_f}(H_{OF,t-t_f}(F_{in}(t)) - F_{in}(t))$$
$$H_{OF,t-t_f}(x) = \begin{cases} x, & \text{if } t < t_f, \\ 3.5\text{ N}, & \text{otherwise.} \end{cases} \tag{3.105}$$

Modeling the ALTAVs and then obtaining the figures of Tables 3.3 and 3.4 is done by carrying out a series of flight experiments, as described in Ref. [143]. The six degree-of-freedom model in closed loop with multi-loop PID autopilots is used for the simulations. Autopilots and distributed controllers given in Secs. 3.1 and 3.2, and parameters shown in Tables 3.5 and 3.6, as well as the observers (3.73)-(3.75) are discretized (Runge-Kutta) with

Table 3.7 Parameters of simplified ALTAV models.

$\omega_1 = 0.65$ rad/s	$\xi_1 = 0.32$	$\omega_2 = 0.35$ rad/s
$\omega_3 = 0.35$ rad/s	$\xi_2 = 0.8$	$\xi_3 = 0.8$
$\omega_4 = 0.39$ rad/s	$\xi_4 = 0.89$	$\omega_5 = 0.33$ rad/s
$\omega_6 = 0.33$ rad/s	$\xi_5 = 0.9$	$\xi_6 = 0.9$
$\omega_7 = 0.35$ rad/s	$\omega_8 = 0.35$ rad/s	$\xi_7 = 1$
$\xi_8 = 1$	$a_1 = -0.035$	$a_2 = 0.01$

a sampling period of 10 milliseconds. The simplified model in Eq. (3.60), which is derived to provide trajectories sufficiently close to those of each ALTAV $i \in \{2, ..., 8\}$ in the formation is characterized by

$$\alpha_i = [\omega_i \; \xi_i],$$

$$A_i = \begin{bmatrix} A_{di}(\alpha) & 0 & 0 \\ 0 & A_{di}(\alpha) & 0 \\ A_{31i}(\alpha) & A_{32i}(\alpha) & A_{di}(\alpha) \end{bmatrix},$$

$$A_{di}(\alpha) = \begin{bmatrix} 0 & 1 \\ -\omega_i^2 & -2\xi_i\omega_i \end{bmatrix},$$

$$A_{31i}(\alpha) = \begin{bmatrix} 0 & 0 \\ 0 & a_1 \end{bmatrix},$$

$$A_{32i}(\alpha) = \begin{bmatrix} 0 & 0 \\ 0 & a_2 \end{bmatrix},$$

$$B_i(\alpha) = diag(B_{di}(\alpha), B_{di}(\alpha), B_{di}(\alpha)),$$

and

$$B_{di}(\alpha) = [0 \; \omega_i^2],$$

along with parameter values given in Table 3.7.

A single robust DAFD observer is implemented onboard each ALTAV. The observer is designed from a pair of polytopic matrices A_i and B_i that encompass state-space matrices of ALTAVs 2 to 8. Note that, for this specific formation, there is an observer onboard follower vehicle i which is based on the model of vehicle $i-1$. For the purpose of observer design, we can exploit the fact that parameters α_i , $i \in \{2, ..., 8\}$, can be bounded by $\omega^* \pm \widetilde{\omega}^M$ and $\xi^* \pm \widetilde{\xi}^M$, where $\omega^* = 0.35$ and $\xi^* = 0.85$. The maximal deviations are equal to 20% of the nominal values. The observer matrices in Eq. (3.73) are computed with the MATLAB® Robust Control Toolbox™ [144]. A robust estimate is achieved with an attenuation gain $\gamma = 0.8$ and with $\mu_i = 1$. The following matrices have been calculated:

$$A_{F,i} = diag(a_{F,i}, a_{F,i}),$$
$$B_{F,i} = diag(b_{F,i}, b_{F,i}),$$

and

$$C_{F,i} = diag(c_{F,i}, c_{F,i}),$$

where

$$a_{F,i} = \begin{bmatrix} -2.56 & 0.95 \\ -0.14 & -0.84 \end{bmatrix},$$
$$b_{F,i} = 10^{-2}[-3.85\ -0.76],$$

and

$$c_{F,i} = \begin{bmatrix} -89.85 \\ 9.64 \end{bmatrix}.$$

The time window used to compute the RMS error is $\tau = 1$ s. The thresholds, described in Sec. 3.3.5, are selected as

$$J_{th,3} = J_{th,4} = J_{th,5} = 0.12,$$
$$J_{th,6} = J_{th,7} = 0.125,$$
$$J_{th,8} = J_{th,9} = 0.13.$$

Note that threshold $J_{th,i}$ is part of DAFD implemented onboard vehicle i to monitor its immediate predecessor labeled as vehicle $i-1$. The inverse dynamics and derivatives are filtered with first-order, low-pass filters whose time constants are set to $\tau_f = 5$ s and $\tau_d = 0.05$ s, respectively.

The command adaptation algorithm described in Fig. 3.26 is implemented onboard each vehicle as follows. A detector flag for fault occurrence is set to one as soon as

$$||r_{ix}||_\tau = ||C_{ix} z_i||_\tau \geq J_{th,i}$$

or

$$||r_{iy}||_\tau = ||C_{iy} z_i||_\tau \geq J_{th,i},$$

where

$$C_{ix} = [1\ 0]$$

and

$$C_{iy} = [0\ 1].$$

With the observer and command adaptation scheme designed, numerical simulations can be run. The simulations aim at verifying that (1) transients induced by the vehicles flying around the corner of the square trajectory do not trigger false alarms, (2) vehicles 3 and 7 detect the faults on ALTAVs 2 and 6 with the same robust observer, which accounts for parametric uncertainty that arises from the simplified formation model used for the design of the observer, and (3) transients of the formation over $[t_f, \tau + t_f)$ do not result in false alarms in vehicles 4, 5, 8, and 9, where τ is the time needed to perform DAFD. Figure 3.36 shows the residues obtained with the numerical simulations. Residues of significant magnitude appear a few seconds after the fault occurs at $t_f = 150$ seconds. Furthermore, the tracking error caused by the vehicle moving around the corner at $t = 100$ seconds does not trigger false alarms. The HOF actuator faults in vehicles 2 and 6 are detected within 4 and 9 seconds, respectively. In the simulations, residues in x cross the thresholds before those residues in y, and therefore the latter are not shown for brevity.

Trajectories on the $x - y$ plane are shown in Fig. 3.37 for faulty vehicles 2 and 6, and their immediate neighbors. The formation starts moving along the square-type trajectory at time 0. The leader reaches the first corner 100 seconds later. Faults on vehicles 2 and 6 occur around the mid-point of the second leg of the square. Then, vehicles 2 and 6 drift away from the required trajectory. We can notice that the DAFD and command adaptation functions are carried out sufficiently quickly by vehicles 3 and 7 to maintain the required formation geometry despite the loss of vehicles 2 and 6. All the other vehicles stay relatively close to one another along the required trajectory.

Positions of the vehicles on the $x - y$ plane at four time instants are shown in Fig. 3.38. Note that the figure is not to scale. The circles represent the ALTAVs. The dotted line represents the square-type trajectory. The arrow indicates the direction of motion of the leader. The vehicles are shown at times $t = 120$, $t = 160$, $t = 175$, and $t = 260$ seconds. Integrity of the formation composed of all the vehicles, but 2 and 6, is preserved as the anomalous behavior of the two faulty vehicles is detected and acted upon sufficiently quickly. The formation does not, however, maintain its original aspect in terms of relative distances ρ^*_{ij} between adjacent neighbors. The vehicles are close to their maximum allowable speed in straight line and thus cannot reduce the inter-vehicle distances that have been abruptly increased by the detection of faulty vehicles 2 and 6, as can be appreciated from Fig. 3.38 at time $t = 175$ seconds. Nevertheless, a slight crossing

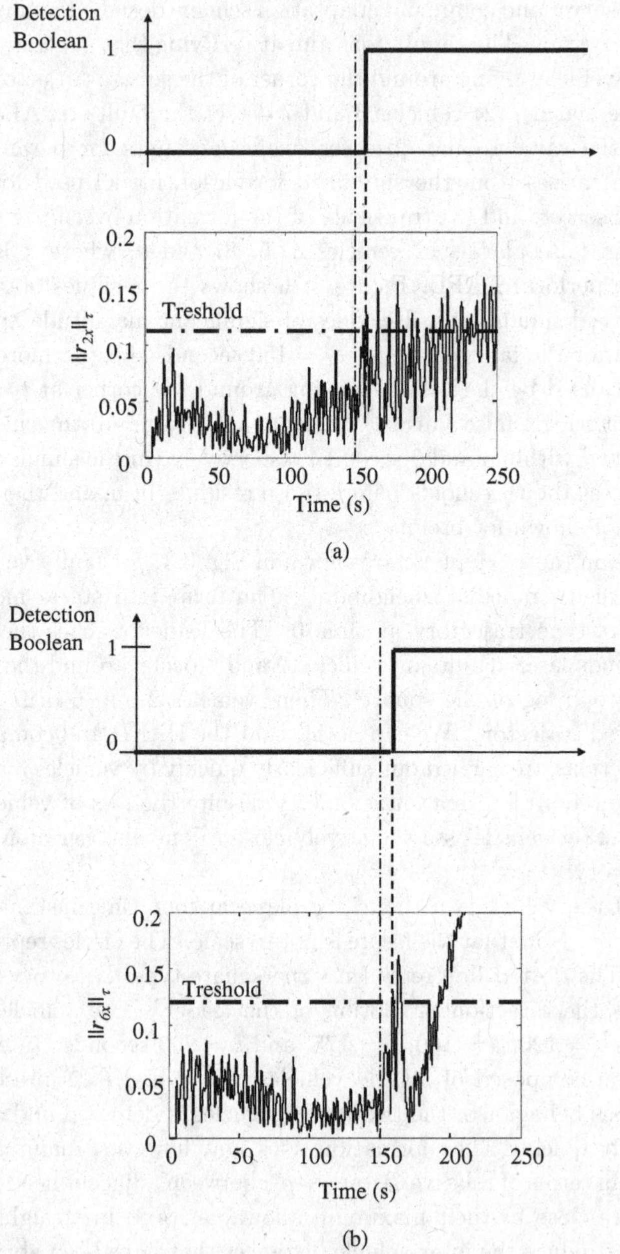

Fig. 3.36 Residues of DAFD observers: (a) onboard vehicle 3 monitoring vehicle 2, and (b) onboard vehicle 7 monitoring vehicle 6.

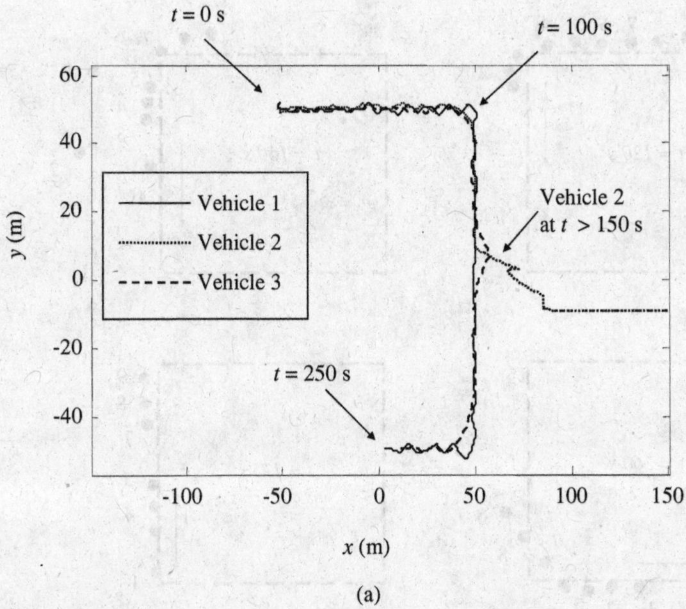

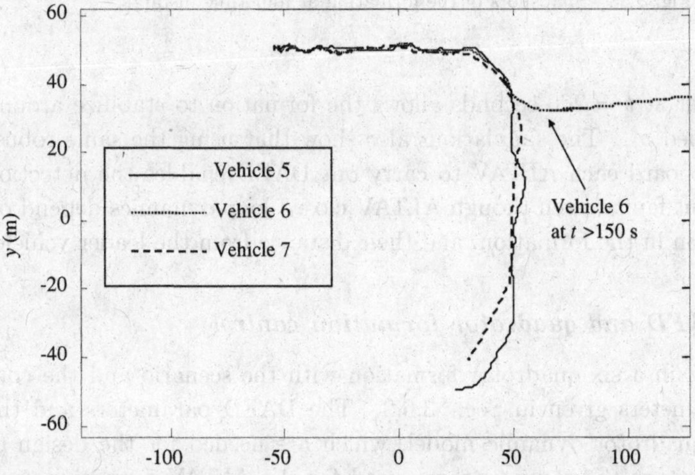

Fig. 3.37 Trajectories of faulty vehicles and immediate neighbors when leader tracks square-type trajectory. (a) Vehicles 1, 2, and 3. (b) Vehicles 5, 6, and 7.

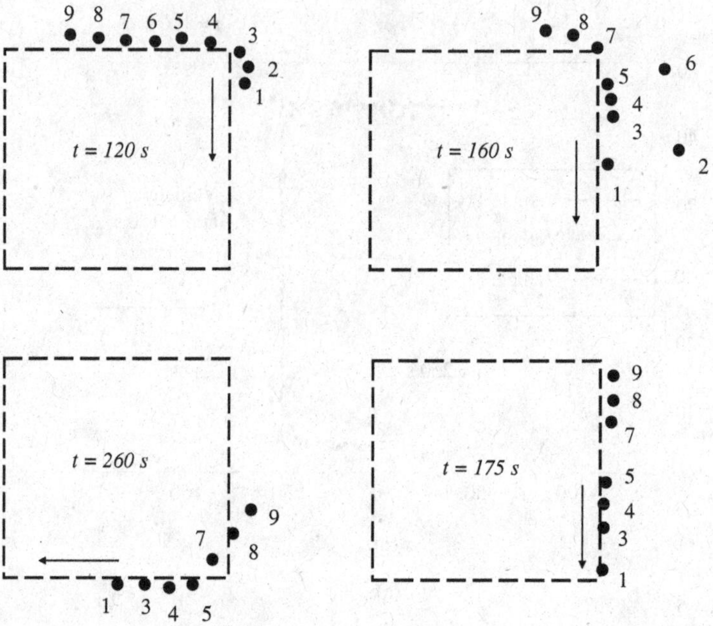

Fig. 3.38 Snapshots of the formation at four time instants.

of the corner at $t = 260$ seconds allows the formation to stabilize around the prescribed ρ^*_{ij}. The simulations also show that using the same robust observer onboard each ALTAV to carry out DAFD enables the detection of concurrent faults, even though ALTAV closed-loop dynamics depend on their position in the formation, and their distance from the leader vehicle.

3.6.4 *DAFD and quadrotor formation control*

Consider again a six-quadrotor formation with the scenario and the controller parameters given in Sec. 3.6.2. The DAFD parameters and the simplified quadrotor dynamic model, which are needed for the design of the observer, are the same as those used for the ALTAV formation, and detailed in Sec. 3.6.3, except for the thresholds. The thresholds are set to a value of 0.015 for the DAFD system of each quadrotor. A faulty vehicle behavior is triggered at a time of 150 seconds into the simulation. The abrupt fault is noticeable on torque τ_θ of vehicle 2, and is shown in Fig. 3.39, where one can notice a sudden change in torque taking place at

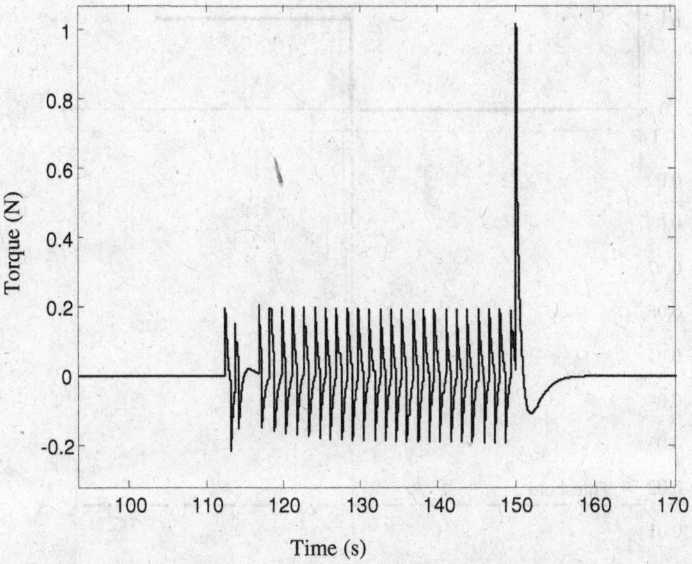

Fig. 3.39 Time history of torque.

time $t = 150$ seconds. The expression for the torque is given by Eq. (3.6). From this equation, it means the faulty behavior of the torque can arise from a fault on motor 2 and/or on motor 4. It is worth recalling, however, that detecting precisely the source of the anomalous behavior of the torque is not of interest here. The health monitoring and adaptation system aims at providing a prompt detection and an appropriate course of action, resulting in coherent formation motion, despite the presence of anomalies. For more details on vehicle numbering in the formation and on the topology of the information flow, the reader is referred to Fig. 3.28.

Figure 3.40 shows the detection boolean and the residue along the x axis, which is calculated onboard quadrotor node 5 to monitor predecessor node 2. The threshold is crossed at about 153 seconds; that is, 3 seconds after the occurrence of the fault. As soon as the residue crosses the threshold, the detector issues a flag with a value of one, meaning there is a fault onboard vehicle 2. From Fig. 3.26, once a fault is detected, a decision is made by node 5 to follow leader node 0, instead of continuing to follow node 2. Furthermore, the formation control law of node 5 no longer commands the vehicle to follow neighboring node 4, but rather to keep the required separation only from the leader vehicle. Figure 3.41 shows the trajectories

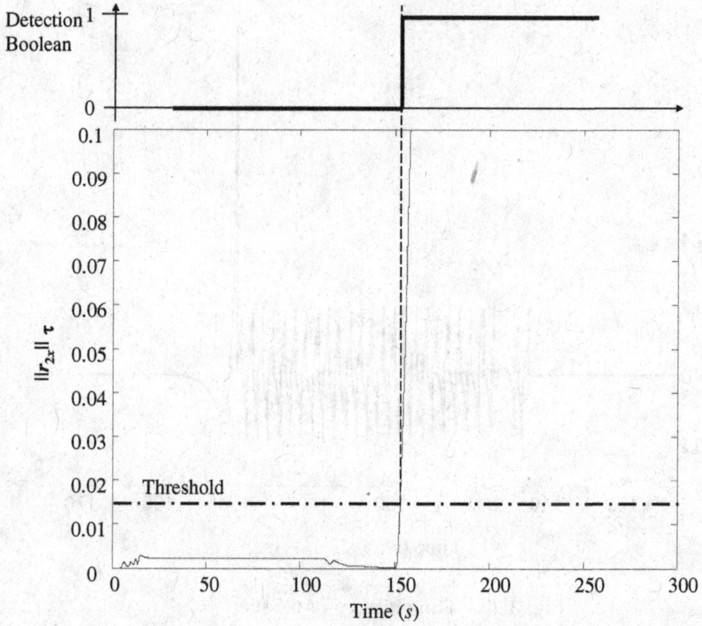

Fig. 3.40 Residues in x and detection boolean for DAFD onboard quadrotor node 5 which monitors quadrotor node 2.

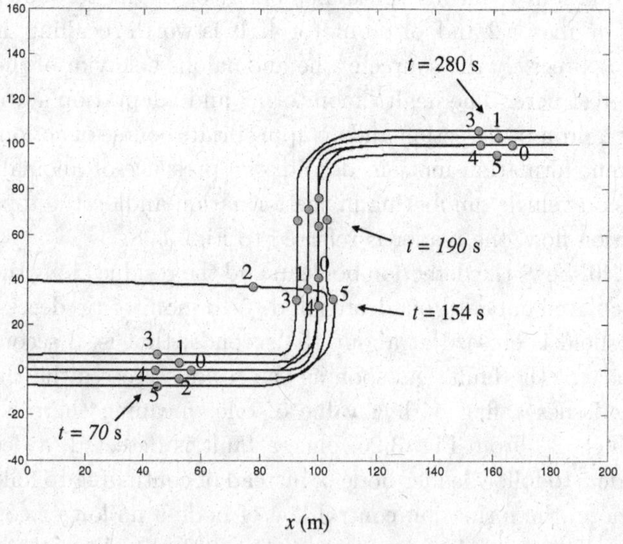

Fig. 3.41 Formation trajectory.

of the 6 vehicles on the $x - y$ plane. The commanded trajectory is the same as that shown in Fig. 3.34. Snapshots of the formation are given at four different time instants in Fig. 3.41. At time $t = 70$ seconds, the formation operates under nominal conditions. Between approximately 100 and 150 seconds, the vehicles execute a circular arc motion. A few seconds after τ_θ begins to exhibit a faulty behavior, node 2 is moving away from the rest of the formation, while its lone successor in the formation, node 5, is not. Onboard node 5, the DAFD and command adaptation systems enable detection of the anomalous behavior of node 2 and adaptation of the formation control law. At $t = 190$ seconds, one can notice the proximity of node 5 to node 4, indicative of transients. After some time, node 5 replaces node 2 in the formation, and the transients have decayed. The formation thus returns to normal operating condition, although with one vehicle missing.

3.6.5 *DNaFD and formation control of ALTAVs*

Controllers (3.2)-(3.4) and (3.58), and DNaFD detector (3.91), (3.98) form a closed-loop system stabilizing a formation of three ALTAVs, $\{1, 2, 3\}$, around a prescribed string shape of regularly spaced and aligned vehicles. Vehicle 1 is the leader, whereas vehicle 3 denotes the last follower vehicle of the string. Neighboring set $\mathcal{N}_i$ is defined, for $i \in \{2, 3\}$, as $\mathcal{N}_i = \{i - 1\}$. The leader of the formation is required to follow a square-type trajectory with side length of 100 m at a speed of $U_0 = 1$ m/s. During mission, either a float fault or a LIP fault, as given in Eq. (3.12), occurs at $t_f = 150$ seconds on one of the motors of ALTAV number 2. Simulation parameters are given in Tables 3.3 and 3.4, whereas controller and filter parameters are provided in Tables 3.5 and 3.6. Furthermore, $k_{s_i} = 0.7$, $k_{p_i} = 0.5$, $a_\gamma = 0.1$ and $a_\phi = 0.1$. Controllers (3.2)-(3.4) and (3.58) are discretized with the Runge-Kutta method at a sampling period of $T_s = 10$ milliseconds. The heading angle estimator in Eqs. (3.92) and the detector in (3.93)-(3.100) are implemented with sampling periods of $T_s = 10$ and $T_N = 10$ milliseconds, respectively. The state-space matrices of the robust observer used in Eqs. (3.92) are computed from the assumption that coefficients C_x and C_y can be expressed as

$$C_x = 0.95 + d_x,$$

and

$$C_y = 0.95 + d_y,$$

where $d_x, d_y \in [-0.2, 0.2]$, yielding the following matrices

$$A_f = \begin{bmatrix} -3.36 \cdot 10^{-7} & 9.24 \cdot 10^{-2} & 0 & 0 \\ -2.98 \cdot 10^{-8} & -2.42 \cdot 10^{-1} & 0 & 0 \\ 0 & 0 & -3.36 \cdot 10^{-7} & 9.24 \cdot 10^{-2} \\ 0 & 0 & -2.98 \cdot 10^{-8} & -2.42 \cdot 10^{-1} \end{bmatrix},$$

$$B_f = \begin{bmatrix} -9.93 \cdot 10^{-1} & 0 & -5.67 \cdot 10^{-1} & 0 \\ 1.18 \cdot 10^{-6} & 0 & -1.45 & 0 \\ 0 & -9.93 \cdot 10^{-1} & 0 & -5.67 \cdot 10^{-1} \\ 0 & 1.18 \cdot 10^{-6} & 0 & -1.45 \end{bmatrix},$$

$$L_f = \begin{bmatrix} 1.06 \cdot 10^{-7} & -4.26 \cdot 10^{-1} & 0 & 0 \\ 0 & 0 & 1.06 \cdot 10^{-7} & -4.26 \cdot 10^{-1} \end{bmatrix}.$$

Once a fault is detected by vehicle l on the immediate predecessor vehicles $l-1$, vehicle l tries establishing an indirect communication link by means of its onboard sensors with vehicle $l-2$, assumed within range. Note that indexes i, j_i, and $k_{j,i}$ used in Sec. 3.4 correspond to vehicles 3, 2, and 1, respectively. The faulty vehicle is j_i, hence $j_i = j_i^F$. The following parameters are used for the detector: $t_{dc} = 15$ seconds, $N_{2,i} = N_{2,k_{j,i}} = 600$, $N_{1,i} = N_{1,k_{j,i}} = 432$, and $v_i = v_{k_{j,i}} = 0.3$, which gives, from simulations, $\alpha \approx 0.8$. Selecting the level of false alarms to $p_F = 0.15$ and solving numerically inequality (3.99) for $n_{s,i}$ gives $p_s(i) = 0.84$.

3.6.5.1 *Simulations exempt from environmental effects*

Suppose a single, non-abrupt fault takes place on an actuator of vehicle number 2. In addition, assume the local FTC-FDD system does not compensate for the fault, which is either LIP or float. In the case of a LIP fault, the actuator is locked at a value of 1.5 N. Under such degraded conditions, the simulations aim at verifying that vehicle 3 is able to detect the actuator fault on vehicle 2 by means of the DNaFD function. The transients induced by the vehicles flying around a corner of the trajectory, the time required for the detection, and the effect of sensor noise are particularly of interest. Sensor characteristics are given in Table 3.4. Figure 3.42 shows time histories for the heading angles of the leader, and follower ALTAVs 2 and 3, starting at a time of 20 seconds into the simulation. For ALTAVs 1 and 3, the heading angles are estimated. The heading angle of ALTAV 2 is used for illustrative purposes only, not to carry out DNaFD.

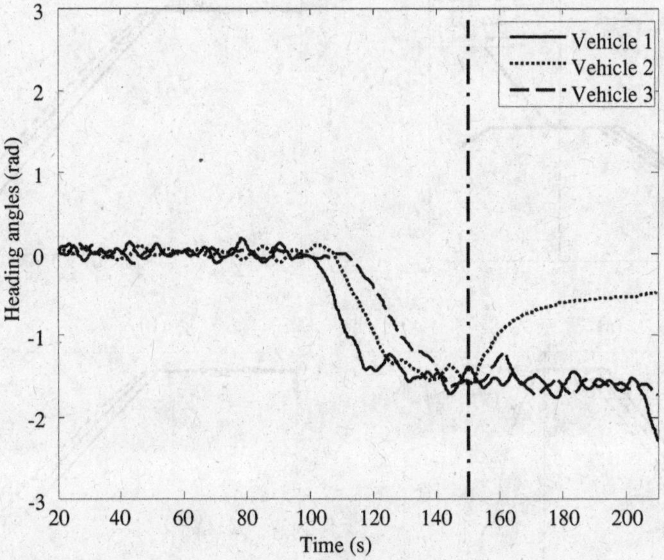

(a)

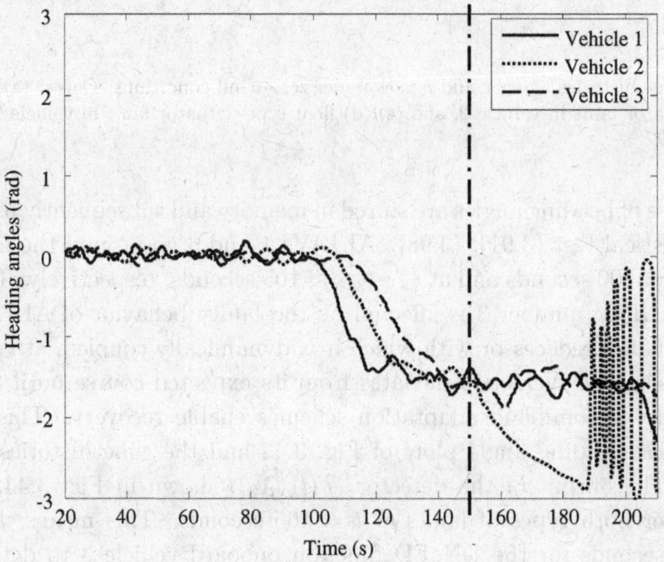

(b)

Fig. 3.42 Estimated heading angles for two types of actuator faults on vehicle 2 at $t_f = 150$ seconds: (a) LIP-type actuator fault, and (b) float-type actuator fault.

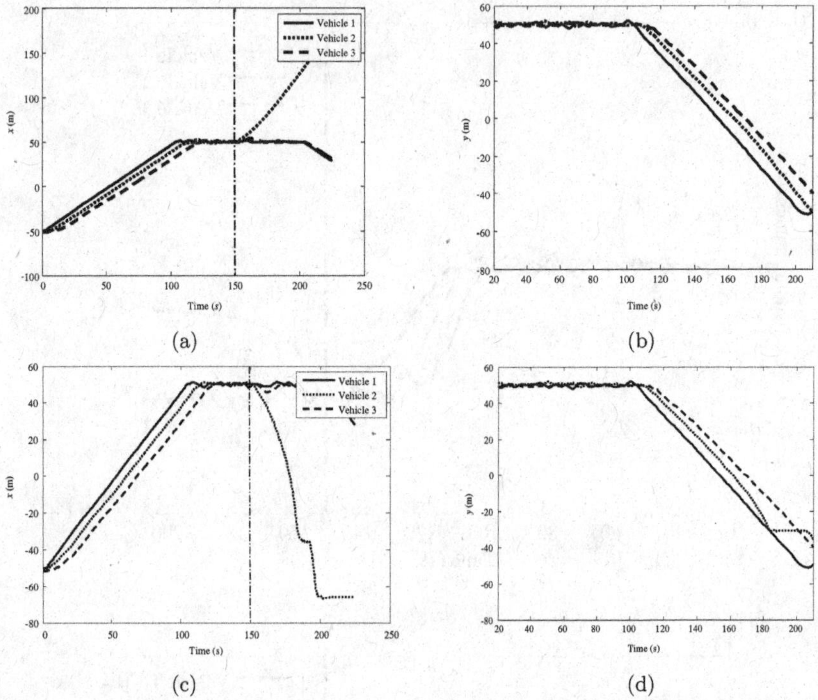

Fig. 3.43 Time histories along x and y axes under zero-wind conditions. Cases: (a),(b) LIP-type actuator fault in vehicle 2, and (c),(d) float-type actuator fault in vehicle 2.

Sequences of heading angles are stored in memory and subsequently used for the statistical test (3.91), (3.98). ALTAVs 1 and 3 go around the first corner at $t_1 = 100$ seconds and at $t_1 + t_{3/1} \approx 105$ seconds, respectively. The motion of ALTAV number 3 is affected by the faulty behavior of ALTAV 2, its immediate predecessor with which it is dynamically coupled, at $t_f + t_{3/2} \approx 160$ seconds. Vehicle 3 deviates from its expected course until the DNaFD and the command adaptation schemes enable recovery. This is clear from the heading angle plots of Fig. 3.42 and the time histories of Fig. 3.43. The output of the detector, $T(t_k, \mathcal{Y}_3)$, shown in Fig. 3.44, is set to one for both types of faults at $t = 165$ seconds. This means that it takes 15 seconds for the DNaFD function onboard vehicle 3 to detect either a LIP or a float fault on an actuator of vehicle 2. In more details, Fig. 3.44 illustrates the detected changes in heading angles of vehicles 1 and 3 by means of $H'(1_{2,3})$ and $H'(3)$, respectively. With a comparison

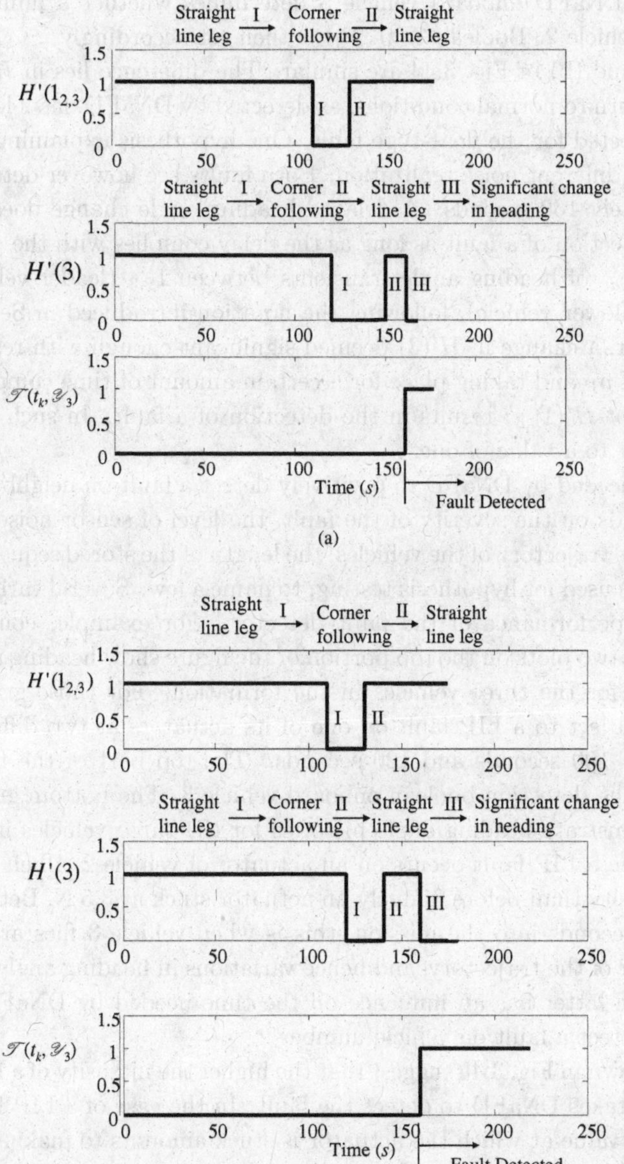

Fig. 3.44 Detection boolean in vehicle 3. The output is set to one when a fault is detected by DNaFD. Two cases are considered: (a) LIP-type fault on an actuator of vehicle 2, and (b) float-type fault on an actuator of vehicle 2.

of sequences, DNaFD onboard vehicle 3 determines whether a fault has occurred on vehicle 2. Boolean $\mathcal{T}(t_k, \mathcal{Y}_3)$ is then set accordingly.

Parts (a) and (b) of Fig. 3.44 are similar. The difference lies in $H'(3)$. For LIP, a return to normal conditions, as detected by DNaFD, lasts longer than that detected for the float-type fault. One hypothesis explaining this difference is a different noise realization. Both faults are however detected at approximately 160 seconds. A delayed heading angle change does not trigger the detection of a fault as long as the delay complies with the propagation time t_{dc} of heading angle transients between $1_{2,3}$ (leader vehicle) and 3 (last follower vehicle), following the notation introduced in Section 3.4.2. However, a change in $H'(3)$ deemed significant enough with respect to the value of p_F and taking place for a certain amount of time compared with a constant $H'(1_{2,3})$ results in the detection of a fault. In such case, $\mathcal{T}(t_k, \mathcal{Y}_3)$ is set to a value of one.

The time needed by DNaFD to positively detect a fault on neighboring vehicles depends on the severity of the fault, the level of sensor noise, the value of p_F, the trajectory of the vehicles, the length of the stored sequences, and the signals used for hypothesis testing, to name a few. Several variables influence the performance of the fault detector. For example, consider Fig. 3.45. The two plots on the top portion of the figure show heading angle time histories for the three vehicles in the formation. For those graphs, vehicle 2 is subject to a LIP fault on one of its actuators at two different time instants: 100 seconds and 120 seconds. The top part of the figure also presents the detection boolean onboard vehicle 3. The bottom graphs of Fig. 3.45 illustrate heading angle histories for the three vehicles in the formation when a LIP fault occurs on an actuator of vehicle 2, albeit with a higher intensity than before, namely an actuator stuck at 3.5 N. Between 100 and 120 seconds into the mission, this is when vehicle 3 flies around the first corner of the trajectory, and hence variations in heading angles are expected. The latter has an influence on the time needed by DNaFD to successfully detect a fault on vehicle number 2.

Results shown in Fig. 3.45 suggest that the higher the intensity of a fault, the shorter it takes DNaFD to detect the fault. In the case of a LIP fault, increasing the value at which the actuator is stuck amounts to making the changes in heading angles more abrupt. Furthermore, results of Figs. 3.42 to 3.45 indicate that a fault can be detected more quickly along a straight leg of the path than along a circular portion of the path. The longer it takes for the detection of a fault, the greater the impact on the integrity of the formation, and hence the higher the probability that the mission will fail.

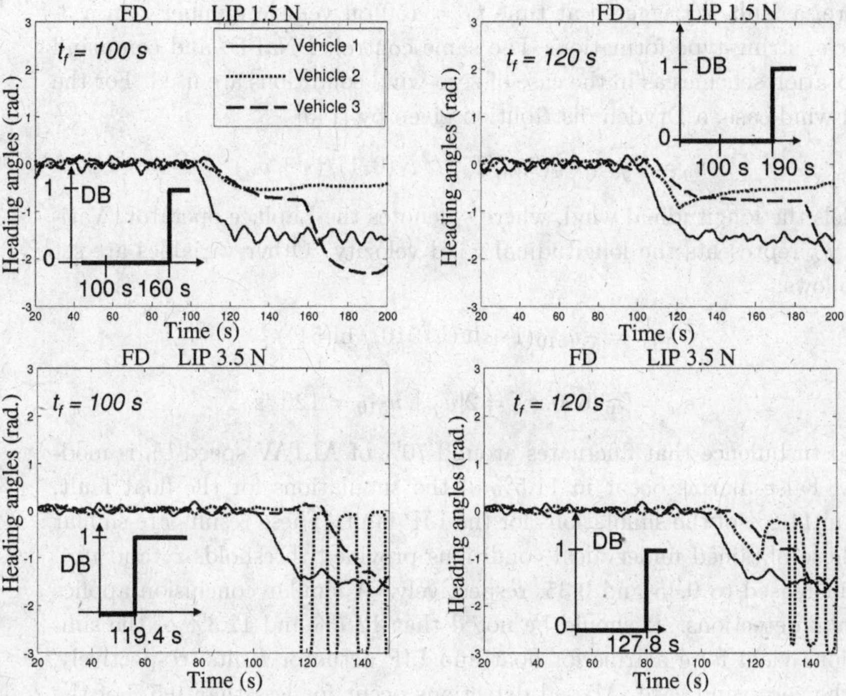

Fig. 3.45 Heading angles and detection boolean. Top: LIP actuator fault at a value of 1.5 N at $t_f = 100$ and $t_f = 120$ seconds. Bottom: LIP actuator fault at a value of 3.5 N at $t_f = 100$ and $t_f = 120$ seconds.

Indeed, any vehicle i in the formation which immediately follows a faulty vehicle takes a path that diverges from its neighbors in the formation until the fault is detected by DNaFD, in which case the healthy vehicle tries to catch up with the preceding healthy vehicles within range. The longer it takes DNaFD onboard vehicle i to detect the presence of a fault on its faulty predecessor, the farther away from the healthy vehicles becomes vehicle i. Furthermore, one can expect an increase in the probabilities of collisions and casualties, and in the demand on the actuators of vehicle i.

3.6.5.2 *Simulations with environmental effects*

Hundreds of simulation runs are carried out for mild wind conditions. Noise realizations vary from one simulation to another. False alarms occur whenever the detection boolean is set to 1 before a fault actually occurs. As before, a fault is triggered at time $t_f = 150$ on vehicle number 2 in a 3-vehicle, string-type formation. The same control, DNaFD, and command adaptation schemes as in the case of zero-wind conditions are used. For the mild wind case, a Dryden distribution given by [145]

$$u_g = u_{gc} + \sigma_u (2a_u/T_s)^{1/2} N(0,1)/(s + a_u)$$

models the longitudinal wind, where s denotes the Laplace operator. Variable u_g represents the longitudinal wind velocity. Other variables are set as follows:

$$u_{gc} = -u_{510}(1 + \ln(h/510)/\ln(51)),$$

$$a_u = U_0/600, \sigma_u = 2|u_{gc}|, u_{510} = 12\text{ft/s}.$$

Wind turbulence that fluctuates around 70% of ALTAV speed U_0 is modeled. False alarms occur in 11.5% of the simulations for the float fault, and in 11.2% of the simulations for the LIP fault. These results are similar to those obtained under ideal conditions provided thresholds v_i and $v_{k_{j,i}}$ are increased to 0.45 and 0.35, respectively. A similar conclusion applies to miss detections. It should be noted that 12.7% and 12.3% of the simulations yield false alarms for float and LIP actuator faults, respectively, for the zero-wind case. Missed detections occur for less than 0.5% of the simulations.

Using threshold values $v_i = v_{k_{j,i}} = 0.3$ for the simulations under mild wind conditions result in a relatively large number of false alarms; namely, false alarms in 40% of the simulations. This suggests using different values for the thresholds depending on wind conditions. For example, one could carry out online estimation of wind parameters, or select thresholds offline for various operating conditions and then store in memory an array of thresholds to be decided by a supervisor for online operation.

Since DNaFD relies on heading angle estimates obtained with a model-based observer (3.92), verifying that parameter uncertainty does not hinder fault detection is important. It was found that levels of false alarms and miss detections remain unchanged when simulations are carried out under zero-wind condition with randomized drag and lift coefficients set to

$$C_x = 0.95 + d_x,$$

$$C_y = 0.95 + d_y,$$

$$C_z = 0.95 + d_z,$$

where d_x, d_y, and d_z are independent random signals that are uniformly distributed over $[-0.2, 0.2]$. This result corroborates the statements made following the design of observer (3.92) that steady-state estimation errors caused by uncertain coefficients with relatively small deviations, here less than or equal to 20% of the nominal values, do not significantly affect the performance of DNaFD.

3.6.6 *Decentralized fault detection for mixed-type, concurrent actuator faults*

To demonstrate that the propagation of multiple, fault-induced transient behaviors within a formation does not result in false alarms and missed detections with DNaFD and DAFD, we again consider a string of nine regularly spaced and aligned ALTAVs, $\{1, ..., 9\}$. The formation of ALTAVs is subject to concurrent faults at $t = t_f$. The formation is required to move along a square-type trajectory, as shown in Fig. 3.35. ALTAV 2 is subject to a non-abrupt float-type fault on one of its four actuators, whereas one actuator of vehicle 6 undergoes an HOF fault ($\theta_{HOF} = \pi/2$). Based on Eq. (3.10) and Fig. 3.9, the HOF is injected in the actuator output F_{4i} of ALTAV 6 as

$$\delta_i = 1_{t-t_f}(H_{OF,t-t_f}(F_{in}(t)) - F_{in}(t))$$
$$H_{OF,t-t_f}(x) = \begin{cases} x, & \text{if } t < t_f, \\ 3.5\text{ N}, & \text{otherwise.} \end{cases} \tag{3.106}$$

where 1_{t-t_f} stands for the Heaviside step function. The low-level controllers, the formation control law, and the DAFD, DNaFD, and command adaptation schemes discussed in the preceding sections for the case of a formation of ALTAVs are implemented in each vehicle with a structure as shown in Fig. 3.27. We assume that the local FTC-FDD systems onboard vehicles 2 and 6 do not provide an acceptable recovery. Trajectories on the $x - y$ plane for faulty vehicles 2 and 6, and for their immediate predecessors and followers, are shown in Figs. 3.46 and 3.47. In Fig. 3.46(a), the time indicated is associated with the position of the leader vehicle. In Figs. 3.46(b),(c), portions of the path are magnified for a better comprehension. In Fig. 3.47, a circle corresponds to a vehicle at a particular time instant. Symbol $t_{i,l}$ means vehicle i at time l. Positions of the vehicles are shown for the following time instants in Fig. 3.47, in units of seconds: 140, 150, 160, 170, 180, 190, 200, 220 and 240. From the figures, it can be seen that the integrity of the formation is relatively well preserved as the

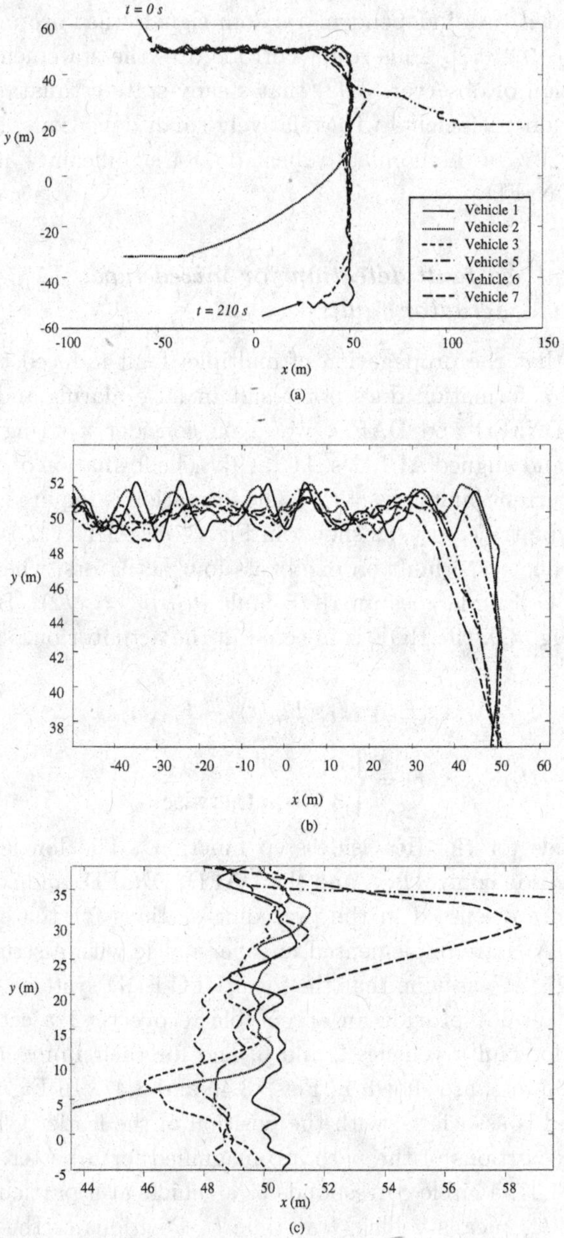

Fig. 3.46 Trajectories for vehicles 1, 2, 3, 5, 6, and 7 on $x - y$ plane: (a) from 0 to 210 seconds, (b) around the first corner, and (c) along the second straight portion of the path.

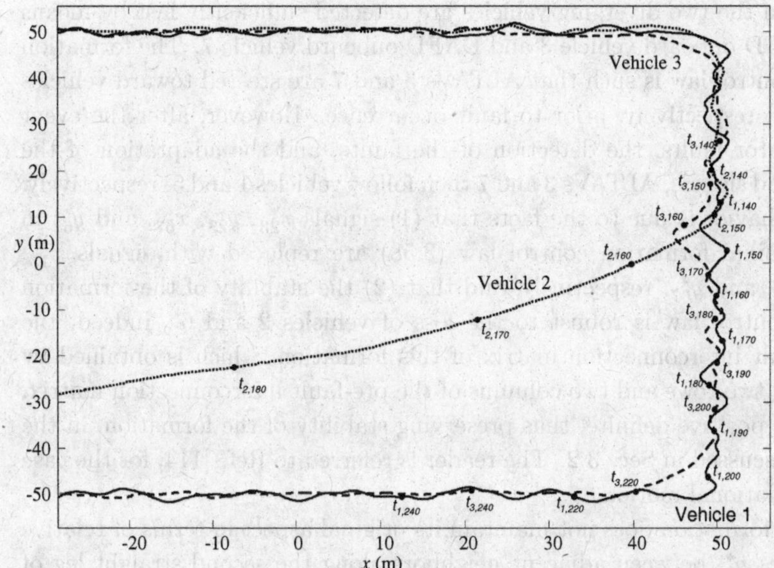

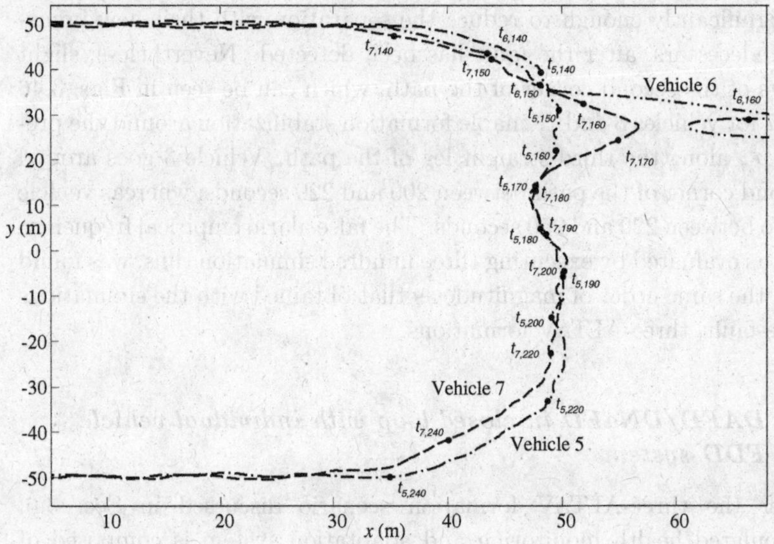

Fig. 3.47 Trajectories and positions at various time instants for (a) vehicles 1, 2, and 3, and (b) vehicles 5, 6, and 7.

faults on the two diverging vehicles are detected sufficiently fast by means of DNaFD onboard vehicle 3 and DAFD onboard vehicle 7. The formation flight control law is such that ALTAVs 3 and 7 are steered toward vehicles 2 and 6, respectively, prior to fault occurrence. However, after the event of actuator faults, the detection of the faults, and the adaptation of the command signals, ALTAVs 3 and 7 then follow vehicles 1 and 5, respectively. This behavior is due to the facts that (1) signals x_{23}^*, y_{23}^*, x_{67}^*, and y_{67}^* in the nominal formation control law (3.58) are replaced with signals x_{13}^*, y_{13}^*, x_{57}^*, and y_{57}^*, respectively, and that (2) the stability of the formation flight control law is robust to the loss of vehicles 2 and 6. Indeed, the post-fault interconnection matrix of this formation, which is obtained by deleting two rows and two columns of the pre-fault interconnection matrix, remains positive definite, thus preserving stability of the formation, in the sense discussed in Sec. 3.2. The reader is referred to Ref. [114] for the case of translational motion.

The formation does not maintain its original aspect in terms of relative distances ρ_{ij}^* between adjacent neighbors along the second straight leg of the path, as shown in Fig. 3.47. ALTAVs are close to their maximum speed in the straight line portion of the trajectory, and thus cannot increase their speed significantly enough to reduce the separation with their new immediate predecessors, after the fault has been detected. Nevertheless, slight crossings of the second corner of the path, which can be seen in Figs. 3.46 and 3.47 for vehicles 3 and 7, enable formation stabilization around the prescribed ρ_{ij}^* along the third straight leg of the path. Vehicle 3 goes around the second corner of the path between 200 and 220 seconds, whereas vehicle 7 does so between 220 and 250 seconds. The false alarm empirical frequency, which was evaluated by executing three hundred simulation runs, was found to be in the same order of magnitude as that obtained with the simulations of single-fault, three-ALTAV formations.

3.6.7 *DAFD/DNaFD in closed loop with individual vehicle FDD system*

Consider the three-ALTAV formation scenario discussed in Sec. 3.6. The simulated health monitoring and adaptation system is composed of DAFD/DNaFD in closed loop with individual vehicle flight control and FDD systems. The simulations rely on the parameters of Sec. 3.6.5. For the purpose of illustration, the nonlinear geometric model-based FDD scheme of Ref. [146] is implemented onboard each ALTAV. A single float-type

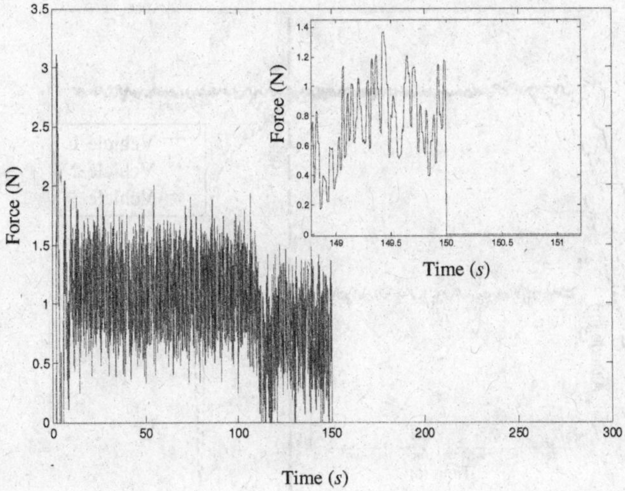

Fig. 3.48 Time history of force generated by faulty motor onboard ALTAV 2.

actuator fault is triggered at $t_f = 150$ seconds onboard ALTAV 2. The faulty signal is shown in Fig. 3.48 over two time scales. At time t_f, the force deployed by the faulty motor falls to 0 N. Such fault results in the heading angle of vehicle 2 behaving as shown in Fig. 3.43. The FDD system of ALTAV 2 detects such actuator fault and then issues a safe-mode command to the vehicle; namely, to fly to a predetermined, safe altitude of 4 m. However, on the $x-y$ plane, vehicle 2 still presents a chaotic trajectory due to the fault. To keep ALTAVs 1 and 3 in a coherent string-type 2-vehicle formation at an altitude of 10 m, the formation resorts to the DNaFD and command adaptation functions available onboard ALTAV number 3. As stated before, DNaFD relies on heading angle estimates and measurements. The main results of the simulations are presented in Fig. 3.49. On the one hand, individual vehicle FDD detects the fault at approximately 180 seconds from the start of the mission. Vehicle 2 lost several meters in altitude before the fault is detected along the z axis, although the vehicle is able to reach the new commanded altitude of 4 m. On the other hand, DNaFD of vehicle 3 detects the faulty behavior of vehicle 2, from the estimated heading angles, at 160 seconds, followed by a command adaptation in all three axes. This is clearly seen on the plots of the $x-y$ plane motion and the altitude of vehicle 3 in Figs. 3.43 and 3.49.

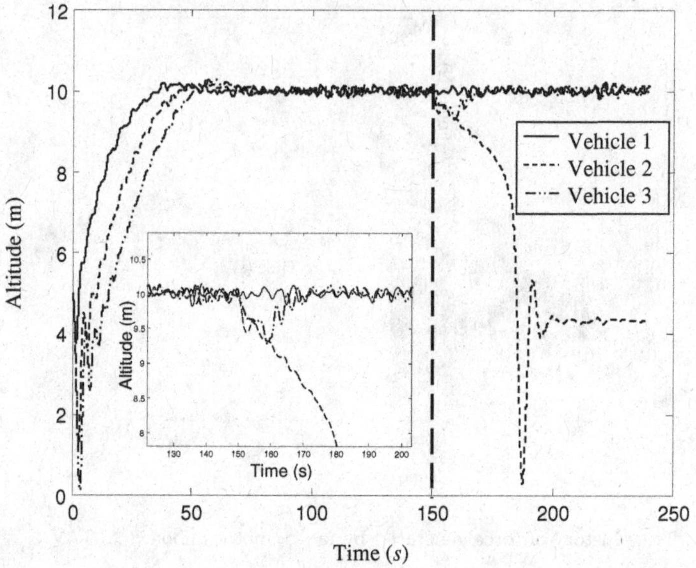

Fig. 3.49 Altitudes obtained with a simulation of the three-vehicle formation.

There is a trade-off between fast detection time and high level of false alarms for both FDD and DAFD/DNaFD, which should be carefully addressed at the design phase in order to have a coherent health monitoring system. Furthermore, the detection time of FDD versus those of DAFD and DNaFD should be tuned prior to the deployment of the vehicles. In the case shown in Fig. 3.49, DNaFD provides detection of the fault faster than the FDD system. However, if faulty vehicle number 2 detects a fault faster than does DNaFD and provides some level of recovery, thus going into a safe mode, this non-abrupt change in the trajectory of the ALTAV may be interpreted as a non-abrupt fault by the DNaFD function implemented in vehicle number 3. This is so as long as the heading angle of vehicle 2 exhibits a sufficient level of change with respect to the heading angle of vehicle 1. Table 3.8 presents various detection times. Detection time is understood as the difference between the time at which a fault is detected by the system and the actual time of fault occurrence. For the FDD system, detection time increases with the value of the threshold. For DNaFD, detection time is constant, as the $x-y$ plane motion of faulty vehicle 2 is unaffected by its safe-mode trajectory, which is confined to controlling the

Table 3.8 Thresholds and detection times.

FDD Threshold	FDD Detection Time	DNaFD Detection Time
0.05	5 s	10 s
0.12	29 s	10 s
0.2	41 s	10 s

motion along the z-axis. It should be noted that the operation of DNaFD is successful for all the values of FDD threshold shown in Table 3.8.

3.6.8 *A note on the digital implementation*

UAS control systems are implemented on digital hardware. The health monitoring and command adaptation functions presented in this book are also intended for implementation on digital hardware. Digital electronic components are getting smaller, of lighter weight, and of lower cost. As mentioned in Ref. [9], "Continued miniaturization is resulting in a migration of capability from larger to smaller platforms." However, the traditional paradigms employed for the design of large aerial systems may no longer be applicable under constraints of mass, size and power faced by small-scale vehicles. Even though computing capabilities are constantly improving, there is still uncertainty about the performance of a system once implemented under constraints. In practice, systems onboard MAVs and SUAVs are subject to constrained digital implementation (CDI) effects. These include finite word length effects, fixed-point computing, control update and sampling rates limited by hardware and subject to variations, signal quantization, and transmission delays among UAS exchanging information via a network or between ground control and UAS. The impact of the CDI effects on the performance of the vehicle, and on the formation, depends on their severity, and on the sensitivity of the systems to such effects. Currently, designers have access to some or all of the following tools when faced with the design of control systems subject to CDI effects: (1) controller design via digital redesign, direct discrete-time design and sampled-data design [139, 147], (2) controller and plant model complexity reduction, (3) several realizations and structures of controllers, (4) discrete-time operators for controller representation and computations [148], (5) encoding/decoding techniques for network communications [149], and (6) selection of sampling and update rates based on various criteria [150].

In this book, the DAFD, DNaFD, and command adaptation functions are modeled and simulated as discrete-time systems, and are therefore ready for digital implementation on the actual platforms. Yet, a statement of caution is in order. The designer should emulate CDI effects and test the performance of the systems by means of extensive simulations, prior to actually deploying the platforms and systems. It is possible that known technological limitations incite the designer to account for constraints (1) in memory, (2) in transmission rate, (3) in the medium used for the communications, (4) in the representation and computations of the models, and (5) in the available sensor information. In the case of DAFD and DNaFD, this means (1) a limited number of thresholds and sequences of signals may be stored in memory onboard the vehicles, (2) information is updated at the rate allowed by the network, which is usually limited, (3) data may be lost during flight depending on the communication medium used and the environment, (4) observer and estimator rely on approximate, reduced-order discrete-time models of actual vehicle dynamics, and (5) information obtained from sensors is converted to digital form, in which case the available digital values are different from the measurements. Finally, the command adaptation function may have to rely on a limited number of gains of approximated value.

Chapter 4

Decision Making and Health Management for Cooperating UAS

Consider a group of unmanned aerial systems accomplishing several tasks concurrently. During the operation, information is shared among the UAS and with ground crew in order to accomplish a certain number of objectives. The crew may intervene at any time, although the UAS team is embedded with systems that enable operations with a relatively high level of autonomy so as to reduce human involvement. Ideally, safe and reliable cooperating unmanned aerial systems perform the tasks needed to achieve a mission on their own, regardless of contingencies of various kinds (see Chapter 1, Sec. 1.3), while limiting risks to the safety of nearby humans and to critical infrastructure.

At the mission level, multi-vehicle cooperation can be understood as a group of vehicles carrying out a sequence of ordered tasks to fulfil the objectives of the mission. At the decision making level, cooperation entails effective ordering of the tasks and use of available resources, typically obtained by means of optimization techniques. At the individual vehicle level, the UAS is guided and controlled in such a way as to provide coherent flight maneuvers appropriate to the tasks at hand, the perceived environment, and the state of the neighboring vehicles (flight maneuvers include hover, loiter, take-off, land, pursuit, and evasion). The cooperating entities rely on passing messages pertaining to their state, the mission, the tasks to be carried out, the flight control commands, the environment (exogenous to the team), and the health of the vehicle flight-critical systems (endogenous to the team).

This chapter studies techniques for leveraging endogenous and exogenous information to provide effective fleet decision making, notably by integrating a health monitoring function into decision making. We argue that more effective team performance can be achieved when a health management system is an integral part of the decision process.

To achieve safe and reliable missions, a number of questions have to be answered in a timely fashion, and solutions have to be obtained online. For example: which vehicle should carry out a specific task? When should a vehicle be replaced by another one to execute a task? In what order should the tasks be performed, given uncertain and often limited endogenous and exogenous information? What should the team do if information on the environment and on some of its members is uncertain or, worse, not available at all? How should the UAS team handle the fact that some entities are not performing as expected? The operating crew may have difficulty answering such questions online, especially if several UAS encounter a variety of contingencies during the operation, if some information is unavailable to the crew, or if the tempo of the mission is high. To obtain answers quickly and translate them into action, we argue in this chapter that concepts of team autonomy should rely on some form of decision making integrated with a health management system. We present CHM techniques integrated with decision making at the team level for the coordinated rendezvous of UAS formations. The architecture we propose is presented in Fig. 4.1.

At the bottom of the hierarchy shown in Fig. 4.1, we have n unmanned systems equipped with guidance, control and FDD-FTC modules. To execute a coordinated rendezvous mission, the vehicles fly as a single collective or as several formations covering different areas of a theater. The health monitoring, adaptation and formation control concepts proposed in Chapter 3 are readily applicable in this context, and in fact serve as baseline teaming functions upon which the functions of CHM and decision making can be developed. Careful integration of the various systems is needed to ensure cohesive overall behavior. Condition monitoring, information management, and inter-vehicle communications ensure information flow between the bottom and the top functions of the hierarchy. Information management collects information, which may come from a sensor network and alternative means of communication, and generates relevant signals to CHM and decision making through information fusion algorithms. Information management handles cases where information is imperfect and missing; for example, when the network is at fault or when vehicles leave the network, are damaged, or lost. Condition monitoring (such as the DAFD and DNaFD systems described in Chapter 3) detects and possibly predicts abnormal conditions using signals from the information management function. The CHM and decision making functions are at the top of the cooperative control hierarchy.

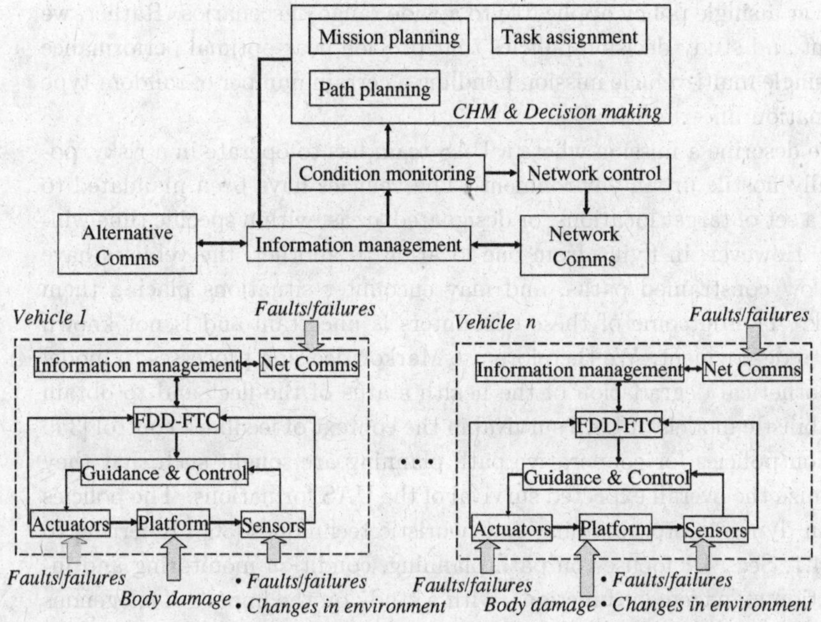

Fig. 4.1 Cooperative health management integrated with decision making for online planning and resource allocation.

Mission planning, task assignment and path planning are cast as problems of optimization and control under uncertainty, with the goal of achieving the most effective team performance in spite of contingencies, partially known environments, and limited information. However, large mathematical problems and combinatorial optimizations give rise to the issue of high dimensionality, an issue which designers must address to ensure that systems can perform in real time. Although not explicit in Fig. 4.1, the functions of CHM and decision making can either be centrally computed on a single processor or distributed over the network of computing nodes. The network control function shown in Fig. 4.1 consists of a feedback policy whose objective is to satisfy mission-specific quality-of-service parameters despite varying delays, unexpected node failures, and packet dropouts. This functionality is outside the scope of this book, however.

In keeping with the no-free-lunch theorem [151], the goal of cooperative decision making that integrates health management is near-optimal performance of a team of UAS in specific scenarios. We do not intend to arrive at a single policy applicable to a wide range of scenarios. Rather, we present and study decision policies that provide near-optimal performance for a single multi-vehicle mission handling a certain number of random-type information uncertainties.

We describe a mission where a UAS team has to operate in a risky, potentially hostile urban environment. The vehicles have been mandated to reach a set of target locations, or designated areas, within specific time windows. However, in flying from one location to another, the vehicles have to follow constrained paths, and may encounter situations placing them at risk. The outcome of these encounters is uncertain and is not known prior to deployment. We therefore use Markov decision processes to model a hypothetical degradation of the health status of the fleet and to obtain probabilistic models of UAS survival in the context of feedback control [35]. Decision policies for cooperative path planning are sought such that they maximize the overall expected survival of the UAS formations. The policies rely on dynamic programming and heuristic techniques [36]. Referring to Fig. 4.1, Sec. 4.1 focuses on path planning, condition monitoring and information management functions, with a study on the impact of exogenous health threats on team decision making. Section 4.2 discusses an additional element of complexity in the problem: the presence of intermittent network communication faults. Indeed, wireless communications are subject to the environment, which may interfere with the signals and block their paths, introducing echoes, noise, and jamming [28]. Failure to share state information among the UAS is likely to reduce the effectiveness of the team. The decision policy requires sharing of information regarding the health state of the vehicles, at the very least to determine whether or not the vehicles have survived a flight within a risky zone. In case there is intermittent loss of communication among the UAS, we propose the design of a recursive Bayesian filter as a health state estimator. Numerical simulations of coordinated rendezvous missions are provided in Sec. 4.3. The distribution of the CHM and decision making functions over the network of computing nodes is briefly discussed in Sec. 4.4.

4.1 Coordinated Rendezvous of UAS Formations

4.1.1 *Context*

Starting from a common base, a team of UAS has been given a mission to reach target areas (waypoints, designated sectors or zones of surveillance) within prescribed time windows with minimum involvement of operating crew. Time windows depend on such factors as location of target areas given UAS capabilities, and optimal flight paths given the level of energy expenditure of the vehicles. Assigning targets or areas of interest to UAS, or formations of UAS, and planning the ordered sequence of areas to be visited may be solved prior to mission, and updated online during the operation. The problem of resource, task, or target assignment has been studied extensively over the past few years. In addressing such problems, one seeks optimal vehicle-target or vehicle-task pairing. Among the variety of techniques that have been investigated, of interest here are stochastic programming [152], genetic algorithms [153], mixed-integer linear programming [154], and the cross-entropy method [155, 156]. In our current scenario, UAS are assumed to be flying at low altitude in a constrained environment. The flight path of a UAS is therefore limited by its surroundings, which consist of buildings, mounts, trees, poles, other vehicles, humans, and so on.

Figure 4.2 illustrates three types of urban pattern [157]: (a) irregular, (b) radial, and (c) grid. Here, we confine our attention to an environment,

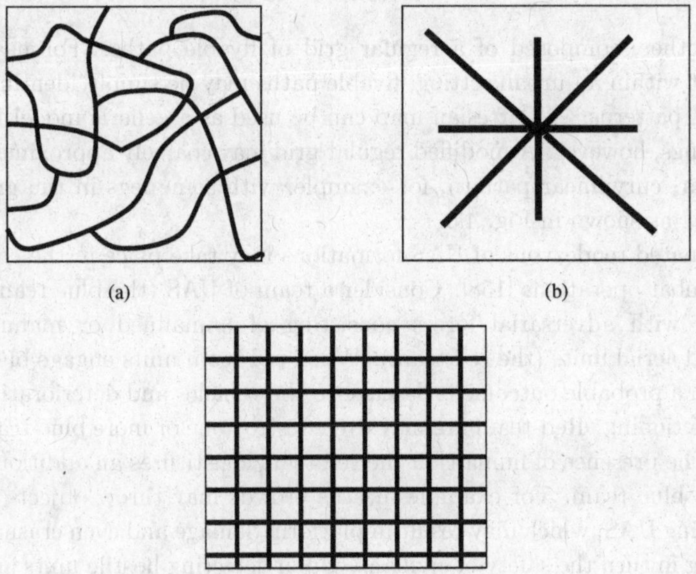

Fig. 4.2 Street patterns.

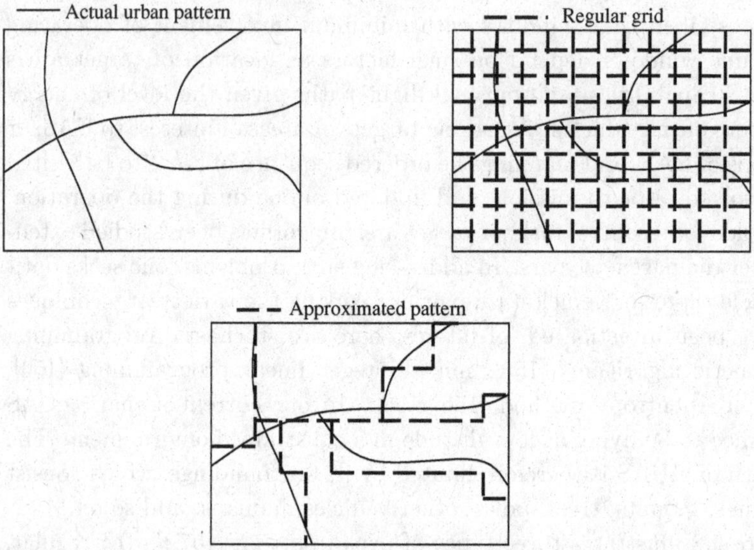

Fig. 4.3 Irregular pattern approximation with edges of the regular grid.

urban or other, composed of a regular grid of flyable paths. For near-earth flight within an urban setting, flyable paths may be simply identified with street patterns. A Cartesian map can be used as a generic model for other settings, however. A modified regular grid may coarsely approximate an irregular, curvilinear pattern, for example, with some legs in the grid unavailable, as shown in Fig. 4.3.

Coordinated rendezvous of UAS formations may take place in the context of combat operations [158]. Consider a team of UAS (the blue team), confronted with adversarial forces consisting of unmanned or manned ground and aerial units (the red team). When red-team units engage blue-team UAS, a probable outcome is damage to the vehicles and deteriorating system functioning. Red-team fire may even destroy one or more blue-team vehicles. The presence of humans in the red team constitutes an additional risk to the blue team. For example, hostile crowds may throw objects at the low-flying UAS, which may result in platform damage and even crashes, threatening in turn the safety of civilians. Upon detecting hostile units and crowds, UAS may choose to alter their flight paths, or the operating crew may impose new restrictions on their pathways. This will increase energy expenditure and reduce the capability of a UAS to complete its mission.

The scope of this book, however, is confined to operations in which UAS do not deploy weapons. In the context we model, the risky, potentially hostile, urban environment consists of areas affected by a disaster, adverse environmental effects, or a dangerous situation, such as contamination, fires, explosions, and so on. Depending on the nature of an event, one may predict a probability of loss of effectiveness in one or more UAS. In this chapter we use probabilities to model the uncertain outcome of situations threatening the operation of the UAS, and hence the success of a mission.

4.1.2 *Related work*

Automating the deployment of multiple UAS operating in a partially known and possibly hostile environment is a challenging task. Systems and crew have access to imperfect and incomplete information, operations are often characterized by opposing forces with conflicting objectives, vehicle dynamics are typically uncertain and vary with time, and the computational and communication capabilities of small-scale UAS are limited. Furthermore, UAS teams may face contingencies such as sensor and actuator malfunctions, platform and onboard component damage, vehicle loss, adverse weather, dynamic obstacles, online mission changes commanded by operating crew, communication network failures, vehicle collisions, and so on.

The deployment of multiple aircraft for rendezvous missions is a field intimately related to command and control of air operations. Material in this chapter is inspired by investigations and original solutions to several problems in the area of air operations, proposals that include probabilistic attrition-type discrete-time models of opposing forces [159–161]. Each party aims at reducing its opponent's capacity for destruction, while maximizing its own survivability. Some researchers rely on game-theoretic lookahead policies to solve this problem [159, 160], proposing one-step and two-step lookahead Nash policies consisting of firing and relocating commands for both teams. To constrain the computing costs, the problem is formulated as a sequence of finite static games. The proposed approach, however, is restricted to a perfect information game, where players have symmetrical roles. Another study addresses air campaign resources in the battlefield, providing task allocation through minimax strategies [162].

One approach to planning air vehicles routes, in particular those in which observation of the state of the opposing team is partial and may entail decoys, lies in a combination of risk-averse control and stochastic dynamic programming [163]. Under the framework of stochastic games, others propose robust feedback control for air vehicles having imperfect information on the ground targets [164–166].

The multi-vehicle mission studied in this section pertains specifically to cooperative path planning or, equivalently, routing management. Path planning and multi-UAV, multi-target assignment can be obtained with Voronoi tessellation and satisficing decision theory [167]. However, the path planner requires perfect knowledge of the theater. Deterministic path planning at the level of individual vehicle flight path typically accounts for collision avoidance to which various constraints can be added to form a mixed-integer linear programming (MILP) problem [168]. Probabilistic path planning that allows a trade-off between shortest path objectives and the mission planner's risk aversion may assume a known probabilistic map of the terrain and the location of threats [169]. Linear programming (LP) relaxation of a MILP optimization of two opposing teams, where one wants to reach a zone protected by defenders, is proposed in Ref. [170] using probabilistic anticipation of the opponents' moves. The suboptimal solution of the LP relaxation is shown to be computationally efficient. Yet, the issue of partial knowledge of the environment and the problem of synchronized vehicle routing in the protected zone are not addressed. Interestingly, the evolution of both teams is modeled as a discrete linear system augmented by a nonlinear attrition function, comparable to the use of Markov decision processes in this book. Although different from our work, cooperative control of UAVs for search missions in partially unknown, risky environments has been modeled and solved by means of dynamic programming in Ref. [171]. References [172, 173] propose centralized and distributed optimal policies for multi-vehicle routing in stochastic, time-varying environments, problems that are closely connected to the one presented in this section. Some researchers have studied problems involving rendezvous, split, and merge of vehicle networks, each vehicle with partial knowledge of the state of the entire network [174, 175]. These algorithms analyze emergent behaviors of multi-vehicle systems and stabilization properties. The schemes are distributed and thus provide robustness to the failures of some of the agents and of communication links.

4.1.3 *Multi-formations*

One of the main features of the cooperative decision making and health management system presented here is its ability to command UAS formations to engage in paths providing, on average, the highest probability of reaching the designated areas, despite exogenous threats to the safety of the vehicles. To do so, the decision policy relies on network communications among the UAS to obtain up-to-date information on the health state of the vehicles, as well as on the solution to an optimization problem formulated such that at any time during mission the UAS team can group into a single formation following a single path, or divide into smaller UAS groups following different paths. The latter are known as multi-formations [158]. The inspiration for the idea of multi-formations comes from the well-known financial principle that a diversified portfolio lowers the risk of losing invested money in the long term. Similarly, enabling multiple formations of a UAS team to engage along different paths within the urban environment is expected to reduce the probability of losing vehicles due to contingencies of various kinds. Intuitively, small formations moving along different paths are more likely to survive firing by adversarial ground units or hostile crowds than a single, large formation [158]. Indeed, imagine standing at a fixed position on the ground facing dozens of UAS moving at a low altitude within sight. A large, tight formation of UAS offers an easier target than a smaller formation of UAS, and the probability of UAS loss is therefore higher. Furthermore, a small formation moving over a hazardous area is likely to suffer fewer losses than a formation of more units. In environments with densely populated pathways, the probability of collisions, and hence of platform damage and loss, increases with the number of UAS. An open air space, on the other hand, offers fewer potential collisions. As another example, suppose the UAS are required to fly between burning buildings. At any moment, windows could break and flames erupt, potentially reaching aerial platforms and affecting performance. The greater the number of UAS near the fire, the greater is the chance that one or more vehicles will be destroyed.

Figure 4.4 illustrates the concept of multi-formations. A team of twelve UAS is shown at four different time instants. The UAS start their flight from the base at time t_1, grouped into two formations. One is comprised of ten vehicles, the other of two. At time t_2, the UAS team flies as two formations of six vehicles over different paths. At t_3, the team is divided into three formations. Finally, at t_4, the vehicles flying under nominal operating conditions have all reached the target area and fly as a single formation of twelve vehicles. Note that the figure does not provide information on the location of the formations. The CHM and decision making functions shown in Fig. 4.1 generate commands for the multi-formations deployment.

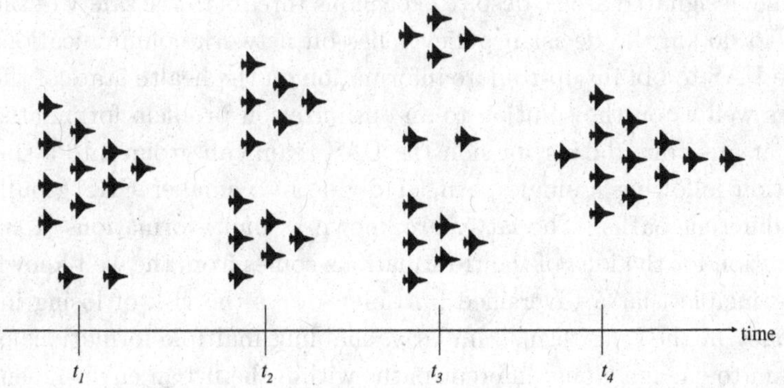

Fig. 4.4 Team of twelve UAS at four different time instants during mission.

4.1.4 *Models*

4.1.4.1 *Cooperating UAS and urban environment*

The urban area, the cooperating UAS and the high-altitude flying surveillance vehicles (SVs) are shown in Fig. 4.5, with top and side views of an idealized urban theater. A team of small-scale, cooperating UAS are deployed in an urban environment. The UAS fly along feasible paths and may be faced with a variety of contingencies. The SVs are located a certain distance away from the urban area. To model routes or legs of the UAS flight path, where contingencies may affect the operation of the UAS, a set of squares are placed in the grid representing the theater, as shown in Fig. 4.5. Following the terminology used in Ref. [158] for the study of combat missions, each square is considered a ground unit. Depending on the context, ground units can be defined as (1) hostile manned or unmanned units that have the capability of engaging the UAS, (2) human individuals or crowds who can throw objects or fire at the UAS, and (3) locations on the grid where UAS are at risk of significant reliability problems (malfunctions, damage, or else) due to exposure to danger and health-threatening situations such as dense environments or adverse phenomena (weather, fires).

A note on the attrition model is in order here. The probability of losing one or more aerial vehicles, due to collision, damage or inter-vehicle communications failure, increases with the size of the UAS formations engaged along so-called perilous legs. These may be routes with a high number of entities, moving obstacles, or dangerous situations. The risk of collision

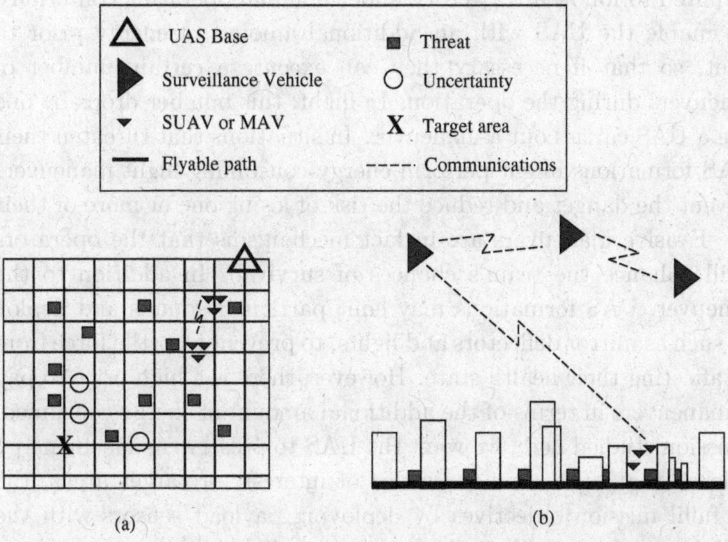

Fig. 4.5 (a) Top and (b) side views of a simplified urban theater.

increases with the number of UAS moving along such routes. Consider a path leg along which several buildings are on fire. Flames may erupt from windows, creating drafts that may reach the aerial vehicles. This path segment would be identified as a perilous leg. The risk of losing one or more vehicles is an increasing function of the number of vehicles taking that leg: if a single gust of fire erupts from a window, and there is only one vehicle taking this route, then the risk of being hit by the gust of fire is smaller than if several vehicles take the same route.

Objectives for combat and civil operations will differ, of course. If the UAS face hostile ground units, a weaponized UAS may engage the ground units. One of the objectives of cooperating UAS involved in a combat mission is to maximize the quantity of available ammunitions once the vehicles are at close range to the targets, which are protected by the ground units. On their way to strategic or tactical targets, UAS may use their weapons against adversarial entities to increase their chances of survival, but at the same time should limit their use to maximize their potential near the target. Reference [158] discusses combat missions in more detail.

For non-combat applications, UAS do not engage opposing forces. In this context, instead of firing ammunition the UAS expend energy in other ways, such as in performing maneuvers to reduce the risk of loss. Besides the energy required to follow a trajectory under nominal operating conditions, operators enable the UAS with an additional amount of energy prior to deployment, so that if necessary they can execute a certain number of extra maneuvers during the operation. In flight, this number drops by one every time a UAS carries out a maneuver. In situations that threaten their safety, UAS formations might perform energy-consuming flight maneuvers to circumvent the danger and reduce the risk of losing one or more of their members. Evasive maneuvers are in fact mechanisms that the operators believe will enhance the team's chances of survival. In addition to the flight maneuver, UAS formations may emit particular sounds and deploy payloads, such as mirror deflectors and lights, to prevent a hostile force from adversely affecting their health state. However, there is a high price to pay for these maneuvers in terms of the additional amount of energy consumed. For the mission studied here, we want the UAS to preserve as much energy as possible so that once near the zones of interest, or target areas, the UAS can fulfil mission objectives by deploying payload/sensors with the required flight maneuvers. Maximizing energy available close to target areas is therefore one objective of the decision policy, yet the conflicting objective of survival on the way to the designated areas of interest may require the UAS formations to use up part or even all of their available energy on evasive maneuvers.

A team of UAS is deployed from a base labeled as B_a. Synchronous arrival at the target areas is expected to maximize the impact of a collective over a single individual. The entire team or a subset of the team is assigned to visit a sequence of targets. The target sequencing problem is described in detail in Refs. [155, 176]. The urban map is assumed known. However, the location and type of some of the threats is typically uncertain. In order to plan their path and determine whether to use specific maneuvers on their way to the targets, SVs and UAS must therefore collect data from onboard and offboard sensors to estimate the location and type of threats affecting their safety. In this book, we limit the number of uncertain threat locations and types. The map is shown in Fig. 4.6. The urban terrain is meshed by a $m_1 \times m_2$ grid denoted as G_{UT}. The nodes in the grid are numbered from 1 to $m_1 m_2$. A node is connected to another node by means of an edge. An edge links two adjacent nodes of the grid. A path is defined as a set of connected edges from B_a to the targets T_i, $i \geq 1$. The time is discretized as $t_k = k \cdot T_s$, $k \in \mathbb{N}^+$, with

$$T_s = \frac{e}{v_\nu}, \tag{4.1}$$

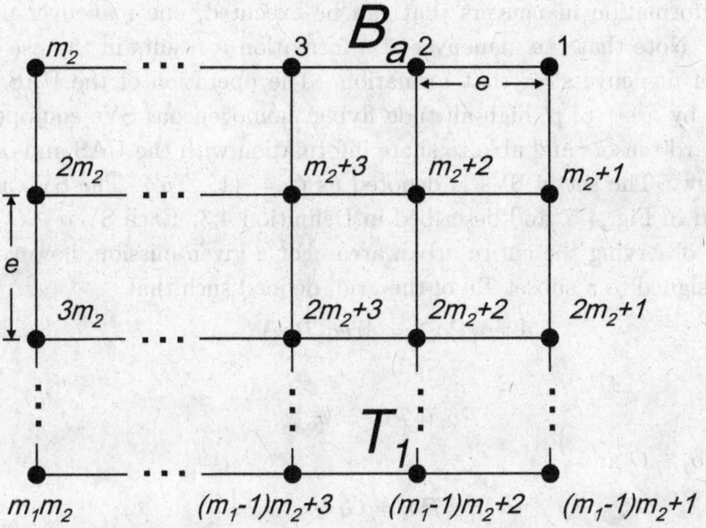

Fig. 4.6 Model of urban terrain as $m_1 \times m_2$ grid G_{UT} with UAS base at node 2 and target area at node $(m_1 - 1)m_2 + 2$.

where e denotes the distance between any two nodes, and v_ν represents the average speed of the UAS along this edge. To ensure synchronous arrival, acceleration of aerial platforms is needed as the UAS maneuver along the edge, and the flight trajectory may have to be reshaped so that the time of flight between any two nodes is fixed to T_s. Thus, at any given time t_k, the UAS are either very near or have arrived at the nodes of G_{UT}. Cooperating UAS and threats are presented in detail in Definitions 4.1 and 4.2, respectively. Assumption 4.1 pertains to the *a priori* knowledge of operating crew on the environment.

Definition 4.1. Cooperating UAS and SVs. At the onset of a mission, p homogeneous formations of UAS are located at base B_a at time t_1. Number p is any positive integer. Any given formation, indexed as ν, is initially composed of n_p UAS. A UAS is instilled with the ability to carry out maneuvers of various kinds, or equivalently has available m_p units of maneuvering energy, at the base. The UAS team, also labeled as B, is able to divide into p distinct formations or into groups of formations ν, a capability labeled as multi-formations. Formations fly along paths from base B_a to target areas T_i. A formation is allowed to maneuver on any given edge of the theater. If p_e formations are engaged along the same edge, there is $p_e \leq p$ formation maneuvers that can be executed; one maneuver per formation. Note that the maneuver of a formation ν results in the use of n_p units of maneuvers for that formation. The operation of the UAS is supported by a set of p' high-altitude flying, homogeneous SVs equipped with onboard sensors and able to share information with the UAS and operating crew. The set of SVs is denoted as $O = \{1, ..., p'\}$. The SVs are exemplified in Fig. 4.5, and described in Definition 4.3. Each SV $o \in O$ is capable of observing the entire urban area. For a given mission, however, SV o is assigned to a subset $\mathcal{P}o$ of the grid, defined such that

$$Area(\mathcal{P}o_i) = Area(\mathcal{P}o_j)$$

and

$$\mathcal{P}o_i \cap \mathcal{P}o_j = \emptyset$$

for all $o_i, o_j \in O$ and

$$\cup_{i=1}^{p'} \mathcal{P}o_i = G_{UT}.$$

Definition 4.2. Perilous edge. Perilous edges refer to edges, or legs, on the grid G_{UT} where lies a threat to the safety of the UAS. The threat, exogenous to the UAS formations, may correspond to (1) hostile manned or unmanned entities that are capable of engaging the UAS, (2) human individuals or crowds who can throw objects or fire at the UAS, or (3) a significant risk of UAS suffering from reliability problems (malfunctions, damage, or else) due to exposure to health-threatening situations, such as dense environments and adverse phenomena (weather, fires). In this book, some of the edges of G_{UT} are perilous edges, while the rest are either safe edges or false-threat edges. The set of threats found in G_{UT} is labeled as R. When p_e formations move along a perilous edge, where lies a threat, there is a possibility of UAS loss. This is as opposed to a safe leg, in which case there is zero probability of UAS loss due to threats exogenous to the team. Note that an endogenous fault onboard a UAS can occur on any edge. A false-threat edge can be simply understood as an area that appears as threatful, but in reality is not. For example, a pacific crowd may seem hostile from the first few observations. However, over time, as more measurements are obtained, the crowd is no longer considered as a threat. A perilous edge offers one of two threat levels to the UAS. On the one hand, a perilous edge may be of a low-level threat type; that is, a threat that may inflict the loss of one UAS of the p_e formations or none at all. On the other hand, a perilous edge of a high-level threat type may result in the loss of several vehicles of the p_e formations.

Assumption 4.1. *A priori* knowledge of environment. We assume that the knowledge of the environment by the team of UAS and by the operating crew is defined by a finite number of possible threat configurations (TC). This number is expressed as γ'. Every TC is labeled as s_γ, with $\gamma \in [1, \gamma']$. A TC corresponds to a particular type associated with every edge of G_{UT}. In case of a partially known environment, it is assumed that the operating crew does not have sufficient information, prior to mission, to determine the actual TC, which is assumed to be within the set $\{s_\gamma, \gamma \in [1, \gamma']\}$.

The $m_1 \times m_2$ grid G_{UT} comprises σ edges, where

$$\sigma = 2m_1m_2 + m_1 + m_2.$$

A TC $s_\gamma, \gamma \in [1, \gamma']$, thus represents, at time t_k, a σ-tuple

$$s_\gamma = \{s_{\gamma,ij}; i \neq j, 1 \leq i \leq m_1, 1 \leq j \leq m_2, s_{\gamma,ij} = s_{\gamma,ji}\},$$

where i and j denote all possible nodes of G_{UT}, and $\text{card}(s_\gamma) = \sigma$. $s_{\gamma,ij}$ characterizes the type of edge (i, j); that is, $s_{\gamma,ij}$ is set to 1 when edge (i, j) is taken as a perilous edge, 2 for a false-threat edge, and 3 when edge (i, j) is considered safe by the UAS team.

4.1.4.2 *Measurements*

UAS and SVs rely in part on their onboard sensors to complete the mission. Definitions 4.3 and 4.4 provide details on the measurement process.

Definition 4.3. Observation. SVs have the capability to observe the entire grid G_{UT}, while UAS can observe the local area. UAS are assumed smaller than SVs, with constrained payload, and hence it makes sense to assume that the UAS have a limited sensing capability. A UAS formation $\nu_l, l \in \{1, ..., p\}$, moving along (i, j), can detect and classify, over $[t_k, t_{k+1})$, the threat lying along that edge. We denote $\mathcal{N}_{Veh}$ as the set of edges that can be sensed by a vehicle $Veh \in \{B, O\}$; in particular, $\mathcal{N}_o = G_{UT}$. Let $\mathcal{N}_{\mathcal{P}o_i}$ be the sensor set of SV o_i restricted to $\mathcal{P}o_i$.

Definition 4.4. Detection and classification. Let $\delta_{ij,k}$ be the distance between any vehicle $Veh \in \{B, O\}$ and a threat, if any, located on edge (i, j) and characterized by $s_{\gamma,ij}$. From the observations, the ability of Veh to achieve a correct detection and classification ($z_{ij,k}^{Veh} = s_{\gamma,ij}$) of the threat is specified by probabilities $p_d^{Veh}(\delta_{ij,k})$ and $p_c^{Veh}(\delta_{ij,k})$, respectively. In practice, $p_d^{Veh}(\delta_{ij,k})$ and $p_c^{Veh}(\delta_{ij,k})$ are decreasing functions of $\delta_{ij,k}$.

The observation variable $Z_{ij,k}^{Veh}$ associated with Veh moving along edge (i,j) can be assigned, at t_k, one of four values, namely $z_{ij,k}^{Veh} = 1, 2, 3$, or nd. The first three values have the same meaning as those attributed to $s_{\gamma,ij}$. A value of 1 indicates a perilous edge, and hence the presence of a health-threatening situation. A value of 2 stands for a false-threat edge, and so there is appearance of a threat, although the segment is safe. A value of 3 corresponds to a safe edge, which is exempt from any exogenous threat to the UAS team. Finally, a value of nd indicates that the detection process has failed. This may occur due to sensor or computing process failure, or due to environmental effects adversely affecting sensor capabilities, to cite a few examples. The measurement model of $Veh \in \{B, O\}$, defined by means of the probability of detection $p_d^{Veh}(\delta_{ij,k})$, the probability of classification $p_c^{Veh}(\delta_{ij,k})$, and the sensor likelihood function $L(z_{ij,k}^{Veh} \mid s_{\gamma,ij})$, can be expressed, at t_k, as

$$
\begin{aligned}
&p(Z_{ij,k}^{Veh} = z_{ij}^{Veh} \mid S_{ij,k} = s_{\gamma,ij}) = p(z_{ij,k}^{Veh} \mid s_{\gamma,ij}) \\
&= \begin{cases} p_d^{Veh}(\delta_{ij,k}) L(z_{ij,k}^{Veh} \mid s_{\gamma,ij}) & \text{if} \quad z_{ij,k}^{Veh} \neq nd, \\ 1 - p_d^{Veh}(\delta_{ij,k}) & \text{if} \quad z_{ij,k}^{Veh} = nd, \end{cases} \qquad (4.2) \\
&L(z_{ij,k}^{Veh} \mid s_{\gamma,ij}) = \begin{cases} p_c^{Veh}(\delta_{ij,k}) & \text{if} \quad z_{ij,k}^{Veh} = s_{\gamma,ij}, \\ \frac{1 - p_c^{Veh}(\delta_{ij,k})}{2} & \text{if} \quad z_{ij,k}^{Veh} \neq s_{\gamma,ij}. \end{cases}
\end{aligned}
$$

Let

$$Z_k^{Veh} = \{Z_{ij,k}^{Veh}; i \neq j, 1 \leq i \leq m_1, 1 \leq j \leq m_2, Z_{ij,k}^{Veh} = Z_{ji,k}^{Veh}\}$$

and

$$Z_k = \{Z_k^{Veh}; Veh \in \{B, O\}\}.$$

For the remainder of the chapter, z_k^{Veh} and z_k are defined similarly to Z_k^{Veh} and Z_k, respectively, with $z_{ij,k}^{Veh}$ replacing $Z_{ij,k}^{Veh}$. Let $\mathcal{Z}_k$ and z_k denote $\{Z_1, ..., Z_k\}$ and $\{z_1, ..., z_k\}$, respectively. In the sequel, S_k stands for a random variable at t_k whose value is taken from the set $\{s_\gamma, \gamma \in [1, \gamma']\}$. Based on the measurement model (4.2), the information state vector, subsequently used by the policy, is represented by the distribution

$$P(S_k = s_\gamma \mid \mathcal{Z}_k = \mathrm{z}_k)$$

expressed over all possible threat configurations $s_\gamma, \gamma \in [1, \gamma']$. Unless indicated otherwise, the short notation $P(S_{\gamma,k} \mid \mathrm{z}_k)$ is used to represent the aforementioned distribution. With a slight abuse of notation, $S_{\gamma,k}$ indicates that, at time t_k, the TC is s_γ. More details on the information state vector are given in Sec. 4.1.8.

4.1.4.3 *Feasible paths*

The requirement of simultaneous arrivals of the UAS at the target areas can be translated into a constraint on the maximum length of the path. In going from the base to the first target area, and from the first target area to the next, and so on, the UAS are constrained to fly along a path with specified minimum and maximum lengths. This allows limiting energy expenditure under nominal operating conditions. Such nominal conditions exclude vehicle maneuvers and actual UAS trajectories, however. The formulation presented here in fact concentrates on the high-level path planning, not on the actual trajectories followed by the UAS. From the base B_a to the first target area T_1, UAS formations are required to follow a path whose length is denoted as $d(B_a, T_1)$. The length of the path is bounded from above and below; namely, $d(B_a, T_1) \in [d_m, d_M]$ where $d_M - d_m = \kappa \cdot e$, $\kappa \in \mathbb{N} \cup \{0\}$. Denote

$$P_0, P_1, ..., P_i, ..., P_\kappa$$

as the sets of feasible paths whose lengths are

$$d_m, d_m + e, ..., d_m + ie, ..., d_M = d_m + \kappa e,$$

respectively. By feasible paths, we mean the paths of length $d(B_a, T_1)$ whose edges do not belong to forbidden zones. Such zones correspond to areas over which the UAS are not allowed to fly.

Define a set E_k, for all $k \in [1, d_m/e + i + 1]$, representing every node of the paths P_i, $i \leq \kappa$, that can be reached by all the formations at t_k; i.e., the set of nodes $n_i \in P_i$, for all $i \leq \kappa$, such that

$$d(B_a, n_i) = (k-1)e,\ k \in [1, \frac{d_m}{e} + i + 1].$$

Simply stated, $d(B_a, n_i)$ represents the distance between base and any node n_i. By definition $E_1 = B_a$ and $E_{d_m/e+i+1} = T_1$. Unless otherwise indicated, symbol T also refers to target T_1. Without loss of generality, the development of a solution to plan the path of UAS formations from B_a to T (or T_1) is the same as that used to plan the paths from T_1 to T_2, and so on. Note that index i relates to paths P_i. Transitions from nodes of E_k to those of E_{k+1} are denoted as $E_{k,k+1}$, and are constrained to paths of P_i that go through nodes of sets E_k and E_{k+1}. The transitions $\mathcal{E}_{k,k+1}$ constitute the edges of a directed acyclic graph, exemplified in Fig. 4.7, and denoted as

$$\mathcal{G}(P_i) = (\mathcal{S}(P_i), \mathcal{E}(P_i)),$$

where $\mathcal{S}(P_i)$ is the set of vertices, or nodes in our case, for paths of P_i and

$$\mathcal{E}(P_i) = \{\mathcal{E}_{k,k+1}, k \in [1, \frac{d_m}{e} + \bar{i}]\}.$$

In the example shown in Fig. 4.7, UAS formations fly from the base at node 4 to a single target area at node 22. The area comprises perilous edges, with threats represented by squares, safe edges indicated by line segments, and false-threat edges denoted by the letter F, as shown in Fig. 4.7(a). The triangle represents the UAS base, and X indicates the location of the target area. The constraint on the total path length, from node 4 to node 22, is set to six legs of length e. For such model of the urban area and constraint on path length, nodes that can be reached by B are shown in Fig. 4.7(b) from singleton E_1 (base) to singleton E_7 (target).

To complete the topic of feasible paths, we need to introduce a few more concepts. We denote the set of all possible distributions of p formations over the nodes of E_k for P_i as

$$C(P_i, B, E_k) = \{c_{1,k}, ..., c_{Cb(p,\eta_k),k}\}, \tag{4.3}$$

where

$$\eta_k = card(E_k)$$

and

$$Cb(p, \eta_k) = \eta_k^p.$$

Cb is the number of permutations with repetitions of p elements taken from a list of η_k elements. For example, with reference to Fig. 4.7(b), the set of all possible distributions for two formations at E_2 is given by

$$C(P_6, B, E_2) = \{(3,3), (3,9), (9,3), (9,9)\},$$

where P_6 is the set of paths of length $6e$, and E_2 is the set of reachable nodes at time t_2. From each distribution $c_{c,k}$ of $C(P_i, B, E_k)$, several permutations with repetitions $c_{d,k+1}$ of nodes of E_{k+1} can be reached at t_{k+1} by the p formations following the authorized transitions from E_k to E_{k+1}. By authorized transitions, we refer to the transitions that occur along paths of P_i. We thus define

$$R_{c_{c,k}} = \{r_{c_{1,k}}, ..., r_{c_{\rho,k}}\}$$

as the set of mappings from $c_{c,k} \in C(P_i, B, E_k)$ to $C(P_i, B, E_{k+1})$ such that

$$r_{c_{d,k}}(c_{c,k}) = c_{d,k+1}, \; d = 1, ..., \rho.$$

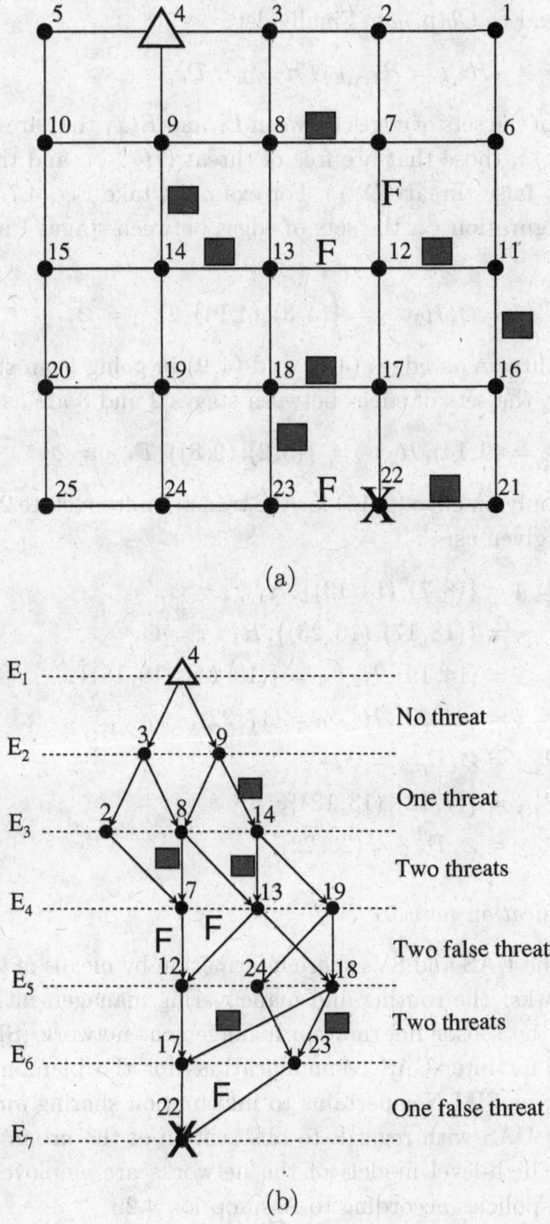

Fig. 4.7 (a) Threat configurations for a single-target mission, and (b) directed acyclic graph representing feasible paths.

Note that ρ is equal to $Cb(p, \eta_k)$. Finally, let

$$\mathcal{H}_{\gamma,k} = \mathcal{H}_{1,\gamma,k} \cup \mathcal{H}_{2,\gamma,k} \cup \mathcal{D}_{\gamma,k}$$

denote the union of the sets of edges between E_k and E_{k+1} that are occupied by a threat ($\mathcal{H}_{1,\gamma,k}$), those that are free of threat ($\mathcal{H}_{2,\gamma,k}$), and those that are occupied by a false threat ($\mathcal{D}_{\gamma,k}$). For example, take Fig. 4.7. For the given threat configuration s_γ, the sets of edges between stages 1 and 2 are given as follows:

$$\mathcal{H}_{1,\gamma,1} = \varnothing, \mathcal{H}_{2,\gamma,1} = \{(4,3),(4,9)\}, \mathcal{D}_{\gamma,1} = \varnothing.$$

So, there are no threats on edges $(4,3)$ and $(4,9)$ in going from stage 1 to stage 2. Similarly, the sets of edges between stages 2 and 3 are as follows:

$$\mathcal{H}_{1,\gamma,2} = (9,14), \mathcal{H}_{2,\gamma,2} = \{(3,2),(9,8)\}, \mathcal{D}_{\gamma,2} = \varnothing.$$

There is a threat only on edge $(9,14)$ for the transition from state 2 to stage 3. Other sets are given as:

$$\begin{aligned}
\mathcal{H}_{1,\gamma,3} &= \{(8,7),(14,13)\}, \mathcal{H}_{1,\gamma,4} = \varnothing,\\
\mathcal{H}_{1,\gamma,5} &= \{(18,17),(18,23)\}, \mathcal{H}_{1,\gamma,6} = \varnothing,\\
\mathcal{H}_{2,\gamma,3} &= (14,19), \mathcal{H}_{2,\gamma,4} = \{(19,24),(19,18)\},\\
\mathcal{H}_{2,\gamma,5} &= (24,23), \mathcal{H}_{2,\gamma,6} = (17,22),\\
\mathcal{D}_{\gamma,2} &= \varnothing, \mathcal{D}_{\gamma,3} = \varnothing,\\
\mathcal{D}_{\gamma,4} &= \{(7,12),(13,12)\},\\
\mathcal{D}_{\gamma,5} &= \varnothing, \mathcal{D}_{\gamma,6} = (23,22).
\end{aligned}$$

4.1.4.4 *Communication network*

We assume that the UAS and SVs share information by means of two communication networks: the routing and maneuvering management network (RMM-Net) and the sensor information management network (SIM-Net). RMM-Net allows for inter-UAS communications for the planning of the flight paths, whereas SIM-Net pertains to information sharing among the SVs and with the UAS with regards to observation of the urban environment. Simplified high-level models of the networks are employed in the design of decision policies according to Assumption 4.2.

Assumption 4.2. Communication networks. First, RMM-Net and SIM-Net are represented by connected, although not necessarily complete, graphs [177]. This means there exists an information path that connects

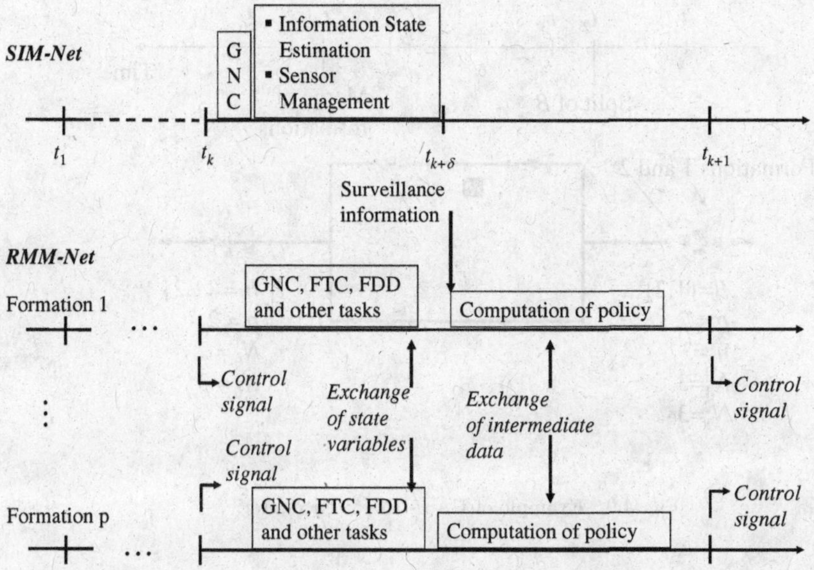

Fig. 4.8 Simplified representation of the communication and computing tasks.

any pair of agents, where agent refers to either a UAS formation or a SV. An information path $\{v_1, v_2, ..., v_d\}$ between v_1 and v_d is composed of pairs of agents $\{v_i, v_{i+1}\}$, $i = 1, \ldots, n-1$, that share information by means of a direct communication link. Second, an operational communication link enables publishing formation locations and health states as well as surveillance information over $[t_k, t_{k+1})$, for all k. Communication and computing tasks for the cooperating UAS and SVs comply with the schedule shown in Fig. 4.8. SVs rely on SIM-Net to provide up-to-date surveillance information to the UAS. RMM-Net is employed by the UAS to relay health status and position, which are essential for the computing of the decision policy. In the figure, computations pertaining to the low-level control tasks and the decision policy are shown. Finally, faulty communication networks are characterized by possible losses of communication links. A loss of communication during interval $[t_k, t_{k+1})$, may isolate one or more agents from the rest of the network, thus resulting in an unconnected network. Such situation may result in a degradation of the performance of the UAS team. Section 4.2 proposes techniques to handle, to some extent, the loss of information resulting from communication network failures.

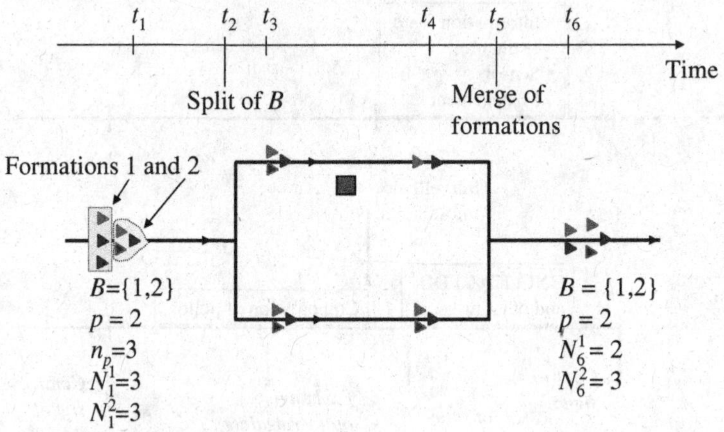

Fig. 4.9 Example of UAS - threat encounter.

4.1.5 *UAS-threat encounters modeled as Markov decision processes*

Each UAS formation $\nu \in B$ is characterized, at time t_k, by a state given by

$$\{N_k^\nu, X_k^\nu\} \in \{0, 1, ..., n_p\} \times \{1, ..., m_1 m_2\}. \tag{4.4}$$

The part of the state labeled as N_k^ν corresponds to the number of operational UAS in formation ν, whereas X_k^ν expresses the node at which the formation is located. The case of an encounter with a low-level threat resulting in a single UAS loss is expressed as

$$N_{k+1}^\nu - N_k^\nu = -1.$$

Figure 4.9 presents an example of an encounter between a UAS formation and a health-threatening situation, or threat. At time instant t_1, B is composed of two formations of three UAS, whose states are denoted by N_1^1 and N_1^2, respectively. B divides into two formations at time t_2. Each formation follows a different path during interval $[t_2, t_5)$, which is the result of the decision policy. One of the formations encounters a threat over $[t_3, t_4]$. This encounter results in the loss of one UAS, and thus in one unit in the value of its state; namely, formation 1 at time t_4 has state $N_4^1 = 2$. At time t_5, the two formations merge into a single formation.

The control signal of formation $\nu \in B$ is given as

$$U_k^\nu = (U_{1,k}^\nu, U_{2,k}^\nu) \in \{j \in \{1, ..., m_1 m_2\}\} \times \{0, 1\} = \mathcal{U}_k^\nu.$$

The first part of the signal, $U^{\nu}_{1,k}$, can be written as $U^{\nu}_{1,k} = X^{\nu}_{k+1}$, and denotes the assigned location of formation ν at time t_{k+1}. The second part of the control signal, $U^{\nu}_{2,k}$, is equal to 0 or 1. Signal $U^{\nu}_{2,k}$ assigned to a value of one represents a maneuver by the entire formation ν over $[t_k, t_{k+1})$, whereas a formation not required to maneuver corresponds to $U^{\nu}_{2,k}$ set to zero.

Let N_k, U_k, $U_{1,k}$, $U_{2,k}$ and $\mathcal{U}_k$ denote the following p-tuples in that order:

$$(N^1_k, ..., N^p_k),$$

$$(U^1_k, ..., U^p_k),$$

$$(U^1_{1,k}, ..., U^p_{1,k}),$$

$$(U^1_{2,k}, ..., U^p_{2,k}),$$

and

$$(\mathcal{U}^1_k, ..., \mathcal{U}^p_k).$$

From the last section, there exists a unique $c_{d,k+1} \in C(P_i, B, E_{k+1})$ such that

$$X_{k+1} = c_{d,k+1}.$$

A threat located on edge (i, j), $i, j \in \{1, ..., m_1m_2\}$, is characterized by a control signal which is a boolean $V_{ij,k} \in \{0, 1\}$. A value of 0 corresponds to a non-threatful situation. Either a hostile entity is not firing at the UAS, or the adverse environmental effect is insignificant at the time of the encounter. The probabilities that a UAS formation loses one of its members is then set to zero. A threat control signal $V_{ij,k}$ set to 1 indicates a health threat to B, or equivalently represents for B a risk of losing one or more UAS along the perilous edge. In short, an active threat menaces the health state of the UAS flying nearby ($V_{ij,k} = 1$), whereas an inactive threat cannot incur any loss to the UAS ($V_{ij,k} = 0$).

Transitions of state variables N^{ν}_k are described by means of Markov decision processes (MDPs) [35]. MDPs model the stochastic nature of attrition dynamics that arise when a UAS formation faces a health threat. Given a threat configuration $s_{\gamma}, \gamma \in [1, \gamma']$, and

$$(U^{\nu}_{1,k}, U^{\nu}_{1,k+1}) = (i, j),$$

let

$$H^{\nu}_{\gamma,k}(i,j) = \begin{cases} H^{\nu}_{1,\gamma,k}(i,j), & \text{when } (i,j) \in \mathcal{H}_{1,\gamma,k}, \\ H^{\nu}_{2,\gamma,k}(i,j), & \text{when } (i,j) \in \mathcal{H}_{2,\gamma,k}, \\ D^{\nu}_{\gamma,k}(i,j), & \text{when } (i,j) \in \mathcal{D}_{\gamma,k}. \end{cases} \tag{4.5}$$

Equation (4.5) indicates the situation faced by a formation ν over time interval $[t_k, t_{k+1})$. There is either a threat, no threat, or a false threat on any given edge of a path. Furthermore, let $H_{\gamma,k}(i,j)$ be a random set whose instance is either $H_{1,\gamma,k}(i,j)$, $H_{2,\gamma,k}(i,j)$ or $D_{\gamma,k}(i,j)$, representing sets of all $H^{\nu}_{1,\gamma,k}(i,j)$, $H^{\nu}_{2,\gamma,k}(i,j)$, and $D^{\nu}_{\gamma,k}(i,j)$, for $\nu = 1, ..., p$, respectively.

Let S indicate that all vehicles of l formations engaged in (i,j) are operational, whereas Vd expresses the fact that at least one of the UAS has been lost as a result of an encounter with a health threat. Then, transition matrices from $\{S, Vd\}$ at t_k to $\{S, Vd\}$ at t_{k+1} can be defined as follows when the two control signals are set to zero,

$$P^{(l)}_b(H_{1,\gamma,k}(i,j), 0, 0) = \begin{bmatrix} 1 & 0 \\ 0 & 1 \end{bmatrix} \begin{matrix} S \\ Vd \end{matrix}, \quad \begin{matrix} S & Vd \end{matrix} \tag{4.6}$$

and as follows for the other three possible combinations of control signals,

$$\begin{aligned} &P^{(l)}_b(H_{1,\gamma,k}(i,j), 0, 1) \\ &\quad = \begin{bmatrix} p^{(l)}_b(H_{1,\gamma,k}(i,j), 0, 1) & 1 - p^{(l)}_b(H_{1,\gamma,k}(i,j), 0, 1) \\ 0 & 1 \end{bmatrix}, \\ &P^{(l)}_b(H_{1,\gamma,k}(i,j), 1, 1) \\ &\quad = \begin{bmatrix} p^{(l)}_b(H_{1,\gamma,k}(i,j), 1, 1) & 1 - p^{(l)}_b(H_{1,\gamma,k}(i,j), 1, 1) \\ 0 & 1 \end{bmatrix}, \\ &P^{(l)}_b(H_{1,\gamma,k}(i,j), 1, 0) = I_2, \\ &P^{(l)}_b(H_{2,\gamma,k}(i,j), ., .) = I_2, \; P^{(l)}_b(D_{\gamma,k}(i,j), ., .) = I_2. \end{aligned} \tag{4.7}$$

Equations (4.6) and (4.7) are MDPs. In Eqs. (4.7),

$$\begin{aligned} &p^{(l)}_b(H_{1,\gamma,k}(i,j), 0, 1) \\ &\quad = P(N^{(l)}_{k+1} = \mathbf{l}_k | N^{(l)}_k = \mathbf{l}_k, H_{\gamma,k}(i,j) = H_{1,\gamma,k}(i,j), \\ &\qquad U^{\nu_1}_{2,k} = 0, ..., U^{\nu_l}_{2,k} = 0, V_{ij,k} = 1), \\ &p^{(l)}_b(H_{1,\gamma,k}(i,j), 1, 1) \\ &\quad = P(N^{(l)}_{k+1} = \mathbf{l}_k | N^{(l)}_{\gamma,k} = \mathbf{l}_k, H_{\gamma,k}(i,j) = H_{1,\gamma,k}(i,j), \\ &\qquad U^{\nu_1}_{2,k} = 1, ..., U^{\nu_l}_{2,k} = 1, V_{ij,k} = 1), \end{aligned}$$

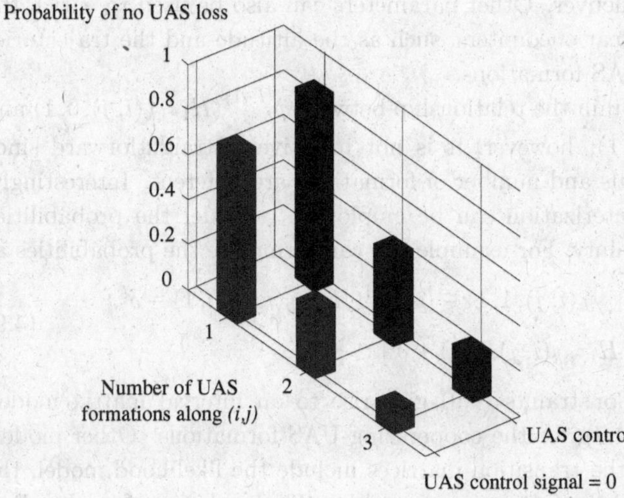

Fig. 4.10 Example of probabilities in UAS-threat encounters from the point of view of the cooperating UAS.

$$N_k^{(l)} = (N_k^{\nu_1}, ..., N_k^{\nu_l}),$$

and

$$\mathbf{l}_k = \{\eta_k^{\nu_1}, ..., \eta_k^{\nu_l}\},$$

where $\eta_k^{\nu_i}$ represents the value that random variable $N_k^{\nu_i}$ may take at t_k; that is, the number of vehicles of formation ν_i at t_k. The modeled probability of survival p_b satisfies the following inequalities

$$\begin{aligned} p_b^{(l)}(H_{1,\gamma,k}(i,j),1,1) &> p_b^{(l)}(H_{1,\gamma,k}(i,j),0,1), \\ p_b^{(l-1)}(H_{1,\gamma,k}(i,j),1,1) &> p_b^{(l)}(H_{1,\gamma,k}(i,j),1,1), \\ p_b^{(l-1)}(H_{1,\gamma,k}(i,j),0,1) &> p_b^{(l)}(H_{1,\gamma,k}(i,j),0,1). \end{aligned} \tag{4.8}$$

Inequalities in (4.8) convey the idea that the risk of vehicle loss increases with the number of formations engaged in (i,j), as stated at the beginning of the section. Figure 4.10 shows the typical probabilistic models employed in this book. The probability of no UAS loss decreases with the number of formations moving along an edge. Furthermore, for a given number of formations near a threat, maneuvers by the formations (control signal set to one) result in a higher probability of survival than if the formations do not execute a maneuver. Other parameters can also be used to enrich the model of UAS-threat encounters, such as the altitude and the trajectories followed by the UAS formations.

We want to obtain the relationship between $p_b^{(l-1)}(H_{1,\gamma,k}(i,j),0,1)$ and $p_b^{(l)}(H_{1,\gamma,k}(i,j),1,1)$; however, it is not intuitively straightforward since both control signals and number of formations are different. Interestingly, an interval characterization can be employed to model the probabilities with some uncertainty. For example, we can formulate the probabilities as

$$p_b^{(l)}(H_{1,\gamma,k}(i,j),1,1) \in [p_b^{(l-1)}(H_{1,\gamma,k}(i,j),0,1) - \delta^l, \\ p_b^{(l-1)}(H_{1,\gamma,k}(i,j),0,1) + \delta^l] \subset [0,1], \tag{4.9}$$

where $\delta^l > 0$. Constraints (4.9) give rise to an interval matrix model, P_b in Eqs. (4.6)-(4.7), for the cooperating UAS formations. Other models of uncertainty in the transition matrices include the likelihood model, the entropy model and the finite-scenario model. The reader is referred to Ref. [178] for more details on such models. The design of the decision policy assumes $p_b^{(l)}(H_{1,\gamma,k}(i,j),1,1)$ is known.

Finally, low- and high-level threats are represented by distribution $P_{b,VehNb}^{(l)}$ over the number of lost UAS, whenever realization of

$$p_b^{(l)}(H_{1,\gamma,k}(i,j),0,1) < 1$$

or

$$p_b^{(l)}(H_{1,\gamma,k}(i,j),1,1) < 1$$

leads to the loss of at least one vehicle. $P_{b,VehNb}^{(l)}$ is a vector $[p_1^{(l)} ... p_{ln_p}^{(l)}]$, where the ith entry is the probability that the effect of the threat results in the loss of i vehicles among the l formations that are engaged along the perilous edge.

4.1.6 *Problem formulation*

Having determined the models, we are now able to find a decision policy for the teaming UAS given as the following sequence

$$\pi_{1,P_i}^u = \{\mu_{1,P_i}, ..., \mu_{N-1,P_i}\}, \ N = d_m + ie.$$

The decision policy provides coordinated rendezvous with maximum capability at the target areas, expressed in terms of maneuvering energy, despite (1) an adversarial, risky, or threatful environment, and (2) imperfect information about the TC. The imperfect information on the TC is represented by means of a probability of having a certain threat configuration (threat location and type) given a measurement variable, namely $P(S_{\gamma,k} \mid z_k)$. With $i \in [0, \kappa]$, $\gamma \in [1, \gamma']$ and $r_{c_{d,k}} \in R_{c_{c,k}}$, μ_{k,P_i} is a function from

$$\{N_k, C(P_i, B, E_k), P(S_{\gamma,k} \mid z_k)\}$$

to

$$\mathcal{U}_k(r_{c_{d,k}}, P_i)$$

such that the decision variable U_k is given as follows

$$\begin{aligned} U_k &= \mu_{k,P_i}(N_k, c_{c,k}, P(S_{\gamma,k} \mid z_k)) \\ &= (\mu_{1,k,P_i}(N_k, c_{c,k}, P(S_{\gamma,k} \mid z_k)), \mu_{2,k,P_i}(N_k, c_{c,k}, P(S_{\gamma,k} \mid z_k))). \end{aligned} \tag{4.10}$$

The decision policy in fact assigns nodes, and hence high-level paths, to the UAS formations at each t_k. Furthermore, the policy determines whether performing extra maneuvering is required.

For R, we also consider a decision policy, even though threats may not actually offer organized decision making, as is the case with adverse environmental effects. We use the term decision policy since we consider R to be an uncooperative player that threatens the safety of the UAS. The decision policy for R is given as a sequence of maps

$$\pi^{\nu}_{1,P_i} = \{v_{1,P_i}, ..., v_{N-1,P_i}\}$$

associated with the threats found along the edges making up the feasible paths. We may associate with the threats the following map

$$v_{k,P_i}(\mu_{k,P_i}(N_k, c_{c,k}, P(S_{\gamma,k} \mid z_k)), \mathcal{H}_{\gamma,k}) = V_k, \tag{4.11}$$

which is assumed unknown to B. In Eq. (4.11), V_k represents the set of control laws of the threats ($\mathcal{H}_{1,\gamma,k} \neq \emptyset$). If there is either a false threat or no threat between nodes of E_k and E_{k+1} ($\mathcal{D}_{\gamma,k} \neq \emptyset$ or $\mathcal{H}_{2,\gamma,k} \neq \emptyset$), V_k is undetermined and transition probabilities are obtained by selecting $H_{\gamma,k} = D_{\gamma,k}$ or $H_{\gamma,k} = H_{2,\gamma,k}$ in Eqs. (4.6)-(4.7).

Determining the policies for the threats and the cooperating UAS is done as follows. Threat and UAS policies play the role of minimizer and maximizer, respectively, of the following function

$$\begin{aligned} J(N_1, P_i, S_{\gamma^+,1}) &= E\{\textstyle\sum_{\nu\in B}(\sum_{k=1}^{k=N-1} AU^{\nu}_{2,k} \\ &\quad + \textstyle\sum_{k=2}^{k=N} m_p I(N^{\nu}_k | N^{\nu}_{k-1}) - S^{\nu})\}, \\ I(N^{\nu}_k | N^{\nu}_{k-1}) &= \left| N^{\nu}_k - N^{\nu}_{k-1} \right|, \\ S^{\nu} &= m_p n_p, \end{aligned} \tag{4.12}$$

where

$$A = \begin{cases} 1, \text{if combat mission} \\ n_p, \text{if non-combat mission.} \end{cases}$$

Function J in Eq. (4.12) represents the negative of the expected number of levels of energy, or maneuvering capability, available to B when the multi-formations reach a target. Function J is in fact an indicator of the health status of the UAS formations. Expectation E is computed with respect to the joint distribution of the random variables N_k^ν involved in MDP (4.7). Function J in Eq. (4.12) is best suited for operations involving a large number of targets, or for wave-planning air operations, where formations are cyclically launched [164]. S^ν is the number of levels of energy, or maneuvering capability, of formation ν at time t_1. $S_{\gamma+,1}$ is the TC utilized at t_1 to derive a policy. The method used to obtain the TC is discussed in Sec. 4.1.8. Function I in Eq. (4.12) is equal to the number of UAS lost, in formation ν, following an encounter with a threat. An increase in J occurs when (1) formation ν fires at the threat (combat) or executes a maneuver (non-combat), to increase its survival probability ($U_{2,k}^\nu = 1$), or (2) one or more UAS are lost due to an encounter with a threat along a risky leg of the path. For the ideal case of perfect information available to B, Problem 4.1 is formulated as follows.

Problem 4.1. ***Perfect information****. Assume that MDP parameters in Eqs. (4.6)-(4.7) are time invariant. MDP parameters are computed prior to mission, and may be updated online. Computation of the probabilities used in Eqs. (4.6)-(4.7) involves a quantitative analysis of the data gathered from past experiences* [179]. *Furthermore, suppose that the actual TC, $S_{\gamma+,1}$, is time invariant and known by B prior to mission. The problem is given by Eq. (4.13). For B, policy π_{1,P_i}^{u+} finds paths that minimize the worst cost. This results in the maximum expected number of maneuvers available satisfying a path length constraint $d(B_a, T) \in [d_m, d_M]$. For R, policy $\pi_{1,P_i}^{\nu+}$ assigns whether a perilous edge of a path represents a risk of losing one or more UAS, with objective of maximizing the expected number of UAS lost.*

$$(\pi_{1,P_i}^{u+}, \pi_{1,P_i}^{\nu+}) = \arg \min_{\pi_{1,P_i}^u} \max_{\pi_{1,P_i}^\nu} J(N_1, P_i, S_{\gamma+,1}) \tag{4.13}$$

For the realistic case of incomplete information on the TC, Problem 4.2 is given as follows.

Problem 4.2. ***Partially known environment****. Given the imperfect nature of the information on the threats available to B and O, obtain the* *following distribution*

$$P(S_{\gamma,k} \mid z_k), \ \gamma \in [1, \gamma']$$

for the TC, valid for all $k \in [1, d_m/e + i + 1]$. The distribution is calculated from past observations and from the model of the target detection and identification process given in Eq. (4.2), from a given initial TC estimate $S_{\gamma^+,1}$. Then, obtain the policies that are near-optimal with respect to those obtained under the idealized conditions of Problem 4.1, and that are computationally tractable.

A few remarks are in order for Problems 4.1 and 4.2. From the information structure suggested in Eqs. (4.10) and (4.11), players B and R have a hierarchical role in which R selects the strategy after B. Such formulation works in favor of R. B makes safe and conservative decisions, in the sense that it minimizes the worst-case scenario. The outcome of a two-player (U, V) game always lies between the lower and the upper values of the game, $\underline{J}$ and $\overline{J}$, respectively, and corresponds to a saddle-point equilibrium,

$$J(U^*, V) \leq J(U^*, V^*) \leq J(U, V^*), \tag{4.14}$$

when $\underline{J} = \overline{J}$ [180]. Hence, the order of play yields an upper value equilibrium $\overline{J}$ which, by definition, is conservative from the perspective of B. Such conservativeness results from the information structure imposed by Eqs. (4.10) and (4.11). However, this information structure may not actually hold. For instance, environmental threats are not dependent on the action of B, implying that

$$\upsilon_{k,P_i}(\mathcal{H}_{\gamma,k}) = V_k. \tag{4.15}$$

Furthermore, if we calculate the optimal sequence of actions for the threats, which maximizes UAS loss, any suboptimal sequence of actions by R will play in favor of B, from inequality (4.14).

Problem 4.2 includes the case of a time-varying TC provided the actual TC belongs to the set $\{s_\gamma, \gamma \in [1, \gamma']\}$ introduced in Assumption 4.1. A decision policy for B determines the paths and whether to carry out maneuvers along perilous edges. This information is then transmitted to the low-level GNC of the leader vehicle of a formation, which calculates a trajectory based on its knowledge of the terrain.

MDP parameters are instrumental to derive an efficient decision policy. Significant differences between the stochastic models of UAS-threat encounters and the model which would be obtained from actual data may result in poor teaming UAS performance. One approach to evaluate the robustness of the decision policy is to calculate the sensitivity of a performance metric, such as the number of healthy UAS at the end of the mission, to changes in the transition probabilities. This is done in Sec. 4.3.3.

4.1.7 *Decision policies: perfect information*

We derive decision policies for two cases. We start with the ideal case described in Problem 4.1. We then extend the policy to handle the more practical case of a partially known environment as stated in Problem 4.2.

To solve Problem 4.1, a dynamic programming equation, or DPE, is formulated. The DPE is central to the problem of cooperative path planning and health management. The DPE relies on graph $\mathcal{G}(P_i)$, described in Sec. 4.1.4, and on stochastic models given by Eqs. (4.6)-(4.7). Denote $S_{\gamma^+,1}$ as the TC used in the backward propagation of the dynamic programming equation. Recall that $S_{\gamma^+,1}$ corresponds to the actual TC when the environment is perfectly known by B. The following proposition provides the solution to Problem 4.1.

Proposition 4.1. [158] *Initially, the state of the UAS team is given as*

$$N_1 = m_p n_p \mathbf{1}_p.$$

Let the cost associated with the optimal solution to Problem 4.1, for TC $S_{\gamma^+,1}$, be equal to $V_{1,P_i,\gamma^+}(N_1, B_a)$. Such cost is obtained from the back-propagation of the cost-to-go function V_{j,P_i,γ^+} via DPE [36]

$$V_{N_N,P_i,\gamma^+}(N_N, T) = -\sum_{\nu \in B} S^\nu,$$

$$\begin{aligned} V_{j,P_i,\gamma^+}(N_j, c_{c,j}) = & \min_{\substack{U_j \in \mathcal{U}_j(r_{c_{d,j}}) \\ r_{c_{d,j}} \in R_{c_{c,j}}}} \max_{V_k} (\textstyle\sum_{\nu \in B} AU^\nu_{2,j} \\ & + \underset{N_{j+1}}{E_{\gamma^+}} \{\textstyle\sum_{\nu \in B} m_p I(N^\nu_{j+1} | N^\nu_j) + V_{j+1,P_i,\gamma^+}(N_{j+1}, c_{d,j+1})\}) \end{aligned} \quad (4.16)$$

for all $j \in \{1, ..., N-1\}$, $c_{c,j} \in C(P_i, B, E_j)$, $c_{d,j+1} \in C(P_i, B, E_{j+1})$, and $N_j \in \{0, 1, ..., n_p\}^p$. In Eq. (4.16), symbol $\underset{N_{j+1}}{E_{\gamma^+}}$ represents the expectation with respect to the joint distribution of N_{j+1}, which is computed from MDPs (4.6)-(4.7) by substituting γ^+ for γ. Optimal policies $\left(\pi^{u+}_{1,P_i}, \pi^{\nu+}_{1,P_i}\right)$, with

$$U_k = \mu_{k,P_i}(N_k, c_{c,k}, P(S_{\gamma,k} \mid z_k)),$$

$$V_k = \upsilon_{k,P_i}(\mu_{k,P_i}(N_k, c_{c,k}, P(S_{\gamma,k} \mid \mathrm{z}_k)), \mathcal{H}_{\gamma+,k}),$$

and

$$P(S_{\gamma+,k} \mid \mathrm{z}_k) = 1,$$

are obtained from the min-max argument of the cost-to-go function $V_{j,P_i,\gamma^+}(N_j, c_{c,j})$ *for all* j.

Proof. The proof follows from Ref. [158]. Let

$$V^*_{N_N,P_i,\gamma^+}(N_N, T) = V_{N_N,P_i,\gamma^+}(N_N, T)$$

and

$$\begin{aligned} &V^*_{j,P_i,\gamma^+}(N_j, c_{c,j}) \\ &\quad = \min_{\pi^u_{j,P_i}} \max_{\pi^\nu_{j,P_i}} \underset{N_{j+1}\ldots N_N}{E_{\gamma+}} \{\textstyle\sum_{k=j}^{k=N-1} A\mu_{2,k,P_i}(N_k, c_{c,k}) \\ &\qquad + \textstyle\sum_{\nu\in B}(\sum_{k=j+1}^{k=N} m_p I(N^\nu_k|N^\nu_{k-1}) - S^\nu)\}, \end{aligned} \tag{4.17}$$

where

$$\pi^u_{j,P_i} = \{\mu_{j,P_i}, ..., \mu_{N-1,P_i}\}$$

and

$$\pi^\upsilon_{j,P_i} = \{\upsilon_{j,P_i}, ..., \upsilon_{N-1,P_i}\}.$$

The terminal value function is $V^*_{N_N,P_i,\gamma^+}$ and the jth-step value function is V^*_{j,P_i,γ^+}. By standard induction arguments [36], it can be shown that the optimal value function $V^*_{j,P_i,\gamma^+}(N_j, c_{c,j})$ is generated by the DPE by assuming its existence at step $j+1$, $V^*_{j+1,P_i,\gamma^+}(N_{j+1}, c_{d,j+1})$, where

$$r_{c_{d,k}}(c_{c,k}) = c_{d,k+1}.$$

Express V^*_{j,P_i,γ^+}, for all $c_{c,j} \in C(P_i, B, E_j)$ and $N_j \in \{0, 1, ..., n_p\}^p$ as

$$\begin{aligned} &V^*_{j,P_i,\gamma^+}(N_j, c_{c,j}) \\ &= \min_{\mu_{j,P_i},\pi^u_{j+1,P_i}} \max_{\upsilon_{j,P_i},\pi^\nu_{j+1,P_i}} \underset{N_{j+1}\ldots N_N}{E_{\gamma+}} \{\textstyle\sum_{k=j}^{k=N-1} A\mu_{2,k,P_i}(N_k, c_{c,k}) \\ &+ \textstyle\sum_{\nu\in B}(\sum_{k=j+1}^{k=N} m_p I(N^\nu_k|N^\nu_{k-1}) - S^\nu)\} \\ &= \min_{\mu_{j,P_i}} \max_{\upsilon_{j,P_i}} \underset{N_{j+1}.}{E_{\gamma+}} \{A\mu_{2,j,P_i}(N_j, c_{c,j}) \\ &+ \textstyle\sum_{\nu\in B} m_p I(N^\nu_{j+1}|N^\nu_j) \\ &+ \min_{\pi^u_{j+1,P_i}} \max_{\pi^\nu_{j+1,P_i}} \underset{N_{j+2}\ldots N_N}{E_{\gamma+}} \{\textstyle\sum_{k=j+1}^{k=N-1} A\mu_{2,k,P_i}(N_k, c_{c,k}) \\ &+ \textstyle\sum_{\nu\in B}(\sum_{k=j+2}^{k=N} m_p I(N^\nu_k|N^\nu_{k-1}) - S^\nu\}\}, \end{aligned} \tag{4.18}$$

by virtue of Lemma 1.6.1 in Ref. [36]. By induction, the bottom equality in Eq. (4.18) can be simplified as

$$V^*_{j,P_i\gamma^+}(N_j, c_{c,j}) = \min_{\mu_{j,P_i}} \max_{v_{j,P_i}} (A\mu_{2,j,P_i}(N_j, c_{c,j})$$
$$+\underset{N_{j+1}}{E_{\gamma^+}}\{\textstyle\sum_{\nu\in B} m_p I(N^\nu_{j+1}|N^\nu_j) + V^*_{j+1,P_i,\gamma^+}(N_{j+1}, c_{d,j+1})\})$$
$$= \min_{\substack{U_j \in \mathcal{U}_j(r_{c_{d,j}}) \\ r_{c_{d,j}} \in R_{c_{c,j}}}} \max_{V_k} (\textstyle\sum_{\nu\in B} AU^\nu_{2,j} \tag{4.19}$$
$$+\underset{N_{j+1}}{E_{\gamma^+}}\{\textstyle\sum_{\nu\in B} m_p I(N^\nu_{j+1}|N^\nu_j) + V_{j+1,P_i,\gamma^+}(N_{j+1}, c_{d,j+1})\})$$
$$= V_{j,P_i,\gamma^+}(N_j, c_{c,j}),$$

where $V_{j,P_i,\gamma^+}(N_j, c_{c,j})$ results in the optimal value function. □

Actual control law U_j applied by B is selected randomly if two or more values of U_j minimize the worst-case cost-to-go function at step j. In more details, for $w \geq 2$ values of U_j, written as $\{U_{j,1}, U_{j,2}, ..., U_{j,w}\}$, we set

$$P(U_j = U_{j,w'}) = 1/w$$

for all $w' \in \{1, ..., w\}$. Note that U_j directly enters function J in Eq. (4.12), and thus the DPE, through $U^\nu_{2,j}$. The DPE depends on $U^\nu_{1,j}$ through the expectation, which is computed with respect to the MDPs given by Eqs. (4.6)-(4.7). We detail the latter through a description of the expectation in Eq. (4.16). Let

$$\mathcal{V}_{j+1,P_i,\gamma^+}(N_{j+1}, c_{d,j+1}) = \sum_{\nu\in B} m_p I(N^\nu_{j+1}|N^\nu_j) + V_{j+1,P_i,\gamma^+}(N_{j+1}, c_{d,j+1}),$$

$$H_{\gamma^+,k} = \xi(\mu_{k,P_i}(N_k, c_{c,k}), \mathcal{H}_{\gamma^+,k}),$$

and

$$l_k = (l_{1,k}, ..., l_{\nu,k}) = L(\mu_{k,P_i}(N_k, c_{c,k}), \mathcal{H}_{\gamma^+,k}),$$

where ξ and L are functions expressing the dependency of $H_{\gamma^+,k}$ and l_k on $\mu_{k,P_i}(N_k, c_{c,k})$ and $\mathcal{H}_{\gamma^+,k}$. Then, the expectation in Eq. (4.16) yields

$$\underset{N_{j+1}}{E_{\gamma^+}}\left\{\mathcal{V}_{j+1,P_i,\gamma^+}(N_{j+1}, c_{d,j+1})\right\}$$
$$= \textstyle\sum_{N_{j+1}\in\mathcal{N}(N_j)} P^{(l_{1,k})}_{xy} ... P^{(l_{\nu,k})}_{xy} \mathcal{V}_{j+1,P_i,\gamma^+}(N_{j+1}, c_{d,j+1}),$$
$$P^{(l_{1,k})}_{xy} = \begin{cases} 1 - p^{(l_{i,k})}_b(H_{\gamma^+,k}(i,j), U_{2,k}, V_k) & \\ \qquad\qquad \text{if } H_{\gamma^+,k}(i,j) = H_{1,\gamma^+,k}(i,j), \\ 1 \qquad\qquad \text{if } H_{\gamma^+,k}(i,j) = H_{2,\gamma^+,k}(i,j), \end{cases} \tag{4.20}$$
$$\textstyle\sum^\nu_{i=1} l_{i,k} = p,$$

where $\mathcal{N}(N_j)$ is the set of all reachable states N_{j+1} starting from state N_j for which at most a single UAS is lost per edge with $\mathcal{H}_{1,\gamma^+,k} \neq \emptyset$. The latter applies to the case of a low-level threat. In case of an encounter with a high-level threat, all states for which several UAS (possibly all) may be lost are reachable. It is worth mentioning that the computation of the decision policy is done from the DPE obtained prior to mission. During mission, each formation calculates and executes the policy from the knowledge of state N_j, communicated by means of RMM-Net, and the DPE.

Despite the attractiveness of representing risks by means of Markov decision processes, statistical analysis of data obtained by sensing the environment may result in uncertain transition matrices. Consider model inaccuracies induced, for example, by environmental effects, such as wind turbulence, fires, and storms. One possible representation of uncertainty is the interval matrix model, where MDP in Eq. (4.7) is now given as

$$\begin{aligned} P_b^{(l)}(H_{1,\gamma,k}(i,j),0,\cdot) &= \begin{bmatrix} p_0 \\ 0 \quad 1 \end{bmatrix}, \\ P_b^{(l)}(H_{1,\gamma,k}(\hat{\imath},j),1,\cdot) &= \begin{bmatrix} p_1 \\ 0 \quad 1 \end{bmatrix}, \end{aligned} \tag{4.21}$$

where, for all $f \in \{0,1\}$,

$$\underline{p}_f \leq p_f \leq \overline{p}_f, \tag{4.22}$$

$$p_f^T \mathbf{1} = 1. \tag{4.23}$$

For the interval matrix model, the decision variable V_k may be interpreted as a high-level or a low-level threat. For a low-level threat, the probability that at least one UAS of B fails to pursue its flight reaches a minimum when, for instance, $V_k = 0$. For a high-level threat, the probability of fault/failure occurrence reaches a maximum, which is achieved when $V_k = 1$. Between both cases, namely $V_k = 0$ and $V_k = 1$, a continuum of conditional probabilities represents the set of model uncertainties in which the model of the actual system is expected to lie. Exploiting the results of robust decision making with uncertain MDPs, as proposed in Ref. [178], the inner maximization combined with the expectation in Eq. (4.16) is replaced by

$$\sup_{P \in \mathcal{P}} P^T \mathcal{V}_{j+1,P_i,\gamma^+},$$

where P denotes the joint distribution of N_{j+1}, and $\mathcal{P}$ represents the set of associated uncertain transition matrices obtained by applying Eqs. (4.21) to the product of matrices in Eqs. (4.20).

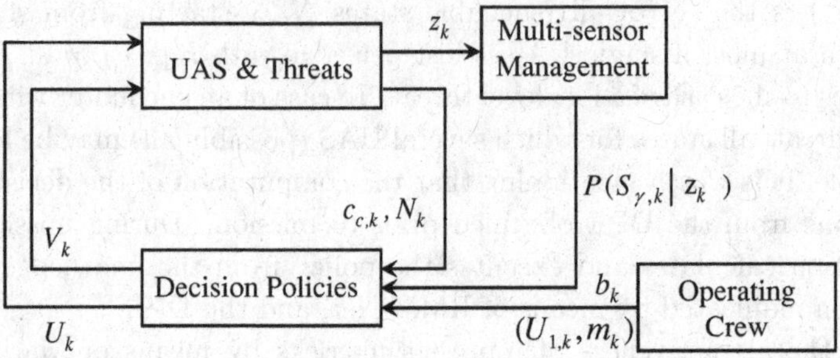

Fig. 4.11 Feedback control interpretation of the cooperative health management and decision making for partially known environment.

4.1.8 *Decision policies: partially known environment*

In case of partial knowledge of the TC, the cooperative decision making and health management of B relies, in part, on an estimation process to solve Problem 4.2. Such process corresponds to the information management block shown in Fig. 4.1. The conditional TC distribution $P(S_{\gamma,k} \mid z_k)$ constitutes an information state upon which relies the decision making of B. The TC estimator in series with the policy enables closing the loop on MDPs (4.6)-(4.7) augmented with measurement model (4.2). This feedback control system interpretation is shown in Fig. 4.11. The plant is in closed loop with the estimator (multi-sensor management) and the decision policies. MDPs are in closed loop with control laws U_k (4.10) and V_k (4.11). At any time during mission, the operating crew can send commands $(U_{1,k}, m_k)$ to the UAS multi-formations by setting boolean b_k. When b_k is set to one, a formation is ordered to maneuver.

Policies π_{1,P_i}^{u+} and $\pi_{1,P_i}^{\nu+}$ are instrumental to the computation of U_k and V_k, respectively. Such policies are expressed in lookup table form and are obtained offline prior to mission. The offline policies are obtained by solving DPE (4.16) as described by Proposition 4.1 for both B and R. The policy for B is updated online based on estimates of the TC. The online update of the decision policy is the subject of this section.

Cooperative health management and decision making of the blue team relies on three components. A policy is obtained prior to mission, based on *a priori* knowledge of the area and of R, even though this knowledge may be imperfect. This policy may be the same policy used to solve Problem 4.1. During the execution of the mission, observation data are used to compute the TC distribution by means of a recursive Bayesian filter (RBF). Since we are in presence of a discrete set, the distribution is a probability mass function over the set of possible TCs. Then, a rollout policy is computed online by each formation in a distributed fashion. The mechanism underlying the rollout policy serves as an improvement step to the nominal decision making calculated offline on the basis of TC $S_{\gamma^+,1}$, which can be selected as the most harmful TC among the possible configurations. Details on the design of the online state estimator and rollout policy are presented.

4.1.8.1 *Information state estimation*

The measurement model (4.2) allows one to express the likelihood $P(z_k \mid S_{\gamma,k})$ as follows

$$P(z_k \mid S_{\gamma,k}) = \prod_{Veh \in \{B,O\}} \prod_{(i,j) \in \mathcal{N}_{Veh}} p(z_{ij,k}^{Veh} \mid s_{\gamma,ij}). \tag{4.24}$$

Without loss of generality, we assume that sensor likelihoods are statistically independent. Then, the *a posteriori* probability, at t_k, of a TC $S_{\gamma,k}$ given observation z_k obtained at t_k, can be expressed as the following RBF

$$P(S_{\gamma,k} \mid \mathrm{z}_k) = \frac{P(z_k \mid S_{\gamma,k}) P(S_{\gamma,k} \mid \mathrm{z}_{k-1})}{\sum_{g=1}^{\gamma'} P(z_k \mid S_{g,k}) P(S_{g,k} \mid \mathrm{z}_{k-1})}, \tag{4.25}$$

where the probability at t_1 is set to $P(S_{\gamma,1})$, as stated in Assumption 4.1. The conditional distribution $P(S_{\gamma,k} \mid \mathrm{z}_k)$ obtained for all $\gamma \in [1, \gamma']$ constitutes the so-called information state.

With the proposed formulation, the likelihood is a function of measurements obtained from sensors onboard every vehicle $Veh \in \{B, O\}$ pertaining to every possible R unit that lies within the sensing set $\mathcal{N}_{Veh}$, as depicted in Fig. 4.12. Several vehicles observe a subset of R. There are therefore multiple measurements for a given entity. Such redundancy may or may not be needed to provide a better estimate of the TC. We could possibly reduce the estimate error of the actual TC by trying to shape, as explained in the sequel, the distribution over the set of TCs. To do so, one seeks to minimize a measure of possible errors by solving a sensor allocation problem. In the figure, the fields of view of SVs, O_1, O_2, and O_3, and that of UAS formation ν_1 intersect to form a sensing set $N_{O_1} \cap N_{O_2} \cap N_{O_3} \cap N_{\nu_1}$ that includes several threats. Ultimately, the allocation problem provides a set of threat-sensor pairs that are retained to compute the TC estimate.

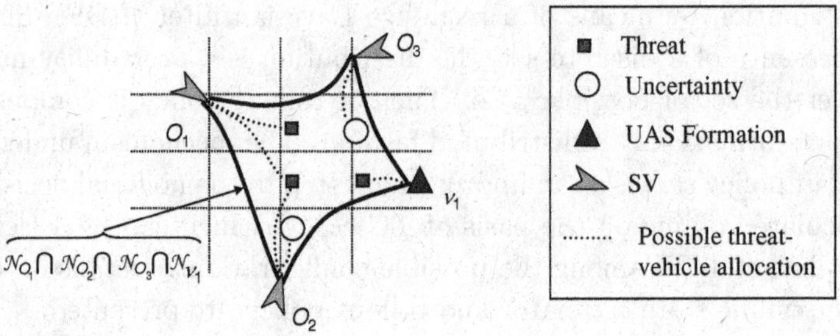

Fig. 4.12 Multi-sensor, multi-target configuration, with possible threat-vehicle allocation.

To solve the sensor allocation problem suggested by Fig. 4.12, namely an information fusion problem, let us introduce the following notation. Denote ε_k, for all t_k, as the list $(\varepsilon_{B,k}, \varepsilon_{O,k})$ of threat-vehicle allocations, where

$$\varepsilon_{B,k} = (\varepsilon_{\nu_1,k}, ..., \varepsilon_{\nu_p,k}),$$

$$\varepsilon_{\nu_i,k} = (\varepsilon_{\nu_i,1,k}, ..., \varepsilon_{\nu_i,\sigma_{\nu_i},k}),$$

$$\varepsilon_{\nu_i,.,k} \in \mathcal{N}_{\nu_i,k},$$

and

$$\sigma_{\nu_i,k} = card(\mathcal{N}_{\nu_i,k})$$

for all $i \in [1, p]$. Furthermore, let $\varepsilon_{o,k}$, $\varepsilon_{o_i,k}$, and $\sigma_{o_i,k}$ be defined similarly. Sensor vector $\Sigma_k \in \{0,1\}^{n_k}$ is associated to ε_k, where

$$n_k = \sum_{i=1}^{p} \sigma_{\nu_i,k} + \sum_{i=1}^{p'} \sigma_{o_i,k}.$$

In practice, sensor vector Σ_k corresponds to all possible threat-vehicle allocations. More precisely, the ith entry of Σ_k is 1 whenever the ith entry of ε_k is retained in the computation of $P(z_k \mid S_{\gamma,k})$. The entry is 0 otherwise. The content of Σ_k indicates which sensor is used to compute

$$P(z_k \mid S_{\gamma,k}) = \prod_{u_k=1}^{n_k} p^{\Sigma_k(u_k)}(z_{u_k} \mid s_{\gamma,ij}), \tag{4.26}$$

where $\Sigma_k(u_k)$ stands for the u_kth entry of Σ_k. z_{u_k} corresponds to $z_{ij,k}^{Veh}$ where $(i,j) \in \mathcal{N}_{Veh,k}$ and Veh represents the SV or UAS associated with the u_kth entry of list ε_k.

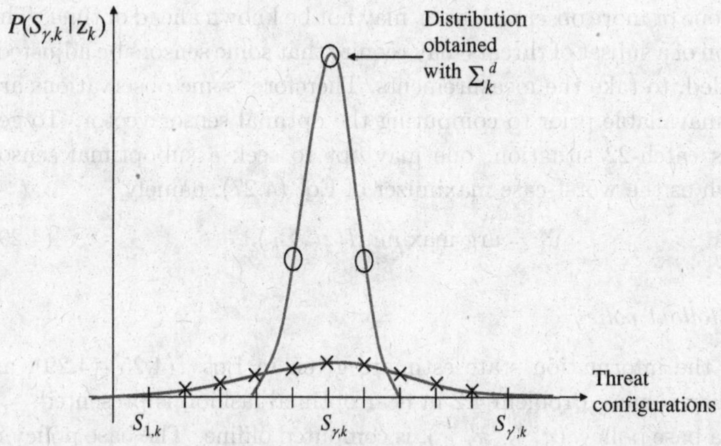

Fig. 4.13 Desired probability mass function (circle) used to estimate the threat configuration.

Sensor vector $\Sigma_k(u_k)$ is determined by minimizing a measure of the degree of error of the estimate of the TC, $\widehat{S}_{\gamma,k}$. As explained in the next section, the maximum *a posteriori* estimator is employed to determine $\widehat{S}_{\gamma,k}$. The ideal estimator would provide a peaky and unimodal probability mass function centered on the actual TC as suggested by Fig. 4.13, where Σ_k^d is the corresponding sensor vector. A flat probability mass function, on the other hand, would lead to larger estimate errors.

Minimizing a measure of the degree of error of the estimate can be accomplished by maximizing the Kullback-Leibler divergence [181]

$$I(z_k, \Sigma_k) = \sum_{g=1}^{\gamma'} P(S_{g,k}|\mathrm{z}_k) \log \frac{P(S_{g,k}|\mathrm{z}_k)}{P(S_{g,k}|\mathrm{z}_{k-1})}$$

$$= \sum_{g=1}^{\gamma'} \frac{P(z_k|S_{g,k})P(S_{g,k}|\mathrm{z}_{k-1}) \log \frac{P(z_k|S_{g,k})}{P(z_k|\mathrm{z}_{k-1})}}{\underbrace{\sum_{g=1}^{\gamma'} P(z_k|S_{g,k})P(S_{g,k}|\mathrm{z}_{k-1})}_{=P(z_k|\mathrm{z}_{k-1})}}, \tag{4.27}$$

where $P(z_k|\mathrm{z}_{k-1})$ depends on Σ_k through $P(z_k \mid S_{\gamma,k})$ given in Eq. (4.26). Ideally, one would seek the sensor vector Σ_k^* that maximizes $I(z_k, \Sigma_k)$; that is,

$$\Sigma_k^* = \arg\max_{\Sigma_k} I(z_k, \Sigma_k). \tag{4.28}$$

However, one or more observations z_k may not be known ahead of time. The observation of a subset of threats may require that some sensors be adjusted, or controlled, to take the measurements. Therefore, some observations are possibly unavailable prior to computing the optimal sensor vector. To get rid of this catch-22 situation, one may opt to seek a suboptimal sensor vector such as the worst-case maximizer of Eq. (4.27), namely,

$$\Sigma_k^* = \arg\max_{\Sigma_k} \min_{z_k} I(z_k, \Sigma_k). \tag{4.29}$$

4.1.8.2 *Rollout policy*

Based on the information state estimate given in Eqs. (4.25)-(4.29), an online policy solving Problem 4.2 in near optimal fashion is presented.

First, a base policy, $(\pi_{1,P_i}^{u+}, \pi_{1,P_i}^{\nu+})$, is computed offline. The base policy is given in Proposition 4.1. To obtain the base policy, the designer may choose set $H_{\gamma^+,k}$ with $S_{\gamma^+,1}$ corresponding to the TC that is the most harmful to B; that is, when each uncertainty is replaced by an active threat. Assuming a distribution over the set of all possible TCs is available, a less conservative approach advocates other types of $S_{\gamma^+,1}$; for instance,

$$\gamma^* = \arg\max_{\gamma\in[1,\gamma']} (P(S_{\gamma,1})). \tag{4.30}$$

Second, for all t_k, $k > 1$, a one-step lookahead policy improvement step is performed over $[t_k, t_{k+1})$ with respect to base policy $(\pi_{1,P_i}^{u+}, \pi_{1,P_i}^{\nu+})$. Such online modification or improvement of the base policy is needed since the estimate of TC $S_{\gamma,k}$ at any time t_k may evolve from that available at time t_1, $S_{\gamma^+,1}$. Then, at time t_k, cost-to-go function V_{k,P_i,γ^+} is approximated over the information state. For that purpose, introduce for all $k > 1$ and all $\gamma \in [1, \gamma']$ the following quantity

$$W_{k,P_i,\gamma}(N_k, c_{c,k}, U_k, V_k) = (\textstyle\sum_{\nu\in B} AU_{2,k}^{\nu} \\ + \underbrace{\underset{N_{k+1}}{E_\gamma} \{\sum_{\nu\in B} m_p I(N_{k+1}^{\nu}|N_k^{\nu}) + \widetilde{V}_{k+1,P_i,,\gamma}(N_{k+1}, c_{d,k+1})\}}_{=W'_{k,P_i,\gamma}(N_k,c_{c,k},U_k,V_k)}), \tag{4.31}$$

where $\widetilde{V}_{k+1,P_i,,\gamma}$ is a cost-to-go function that depends on base policy $(\pi_{k+1,P_i}^{u+}, \pi_{k+1,P_i}^{\nu+})$ of Proposition 4.1 and on the entire set $\{s_\gamma, \gamma \in [1, \gamma']\}$ of possible threat configurations. Calculating this function requires solving DPE (4.16), which is computationally expensive as $\widetilde{V}_{k+1,P_i,,\gamma}$ now depends on three arguments: N_{k+1}, $c_{c,k+1}$ and $\gamma \in [1, \gamma']$. In Eq. (4.16), the TC is fixed to $S_{\gamma^+,k+1}$, the actual environment. Here, the environment is not

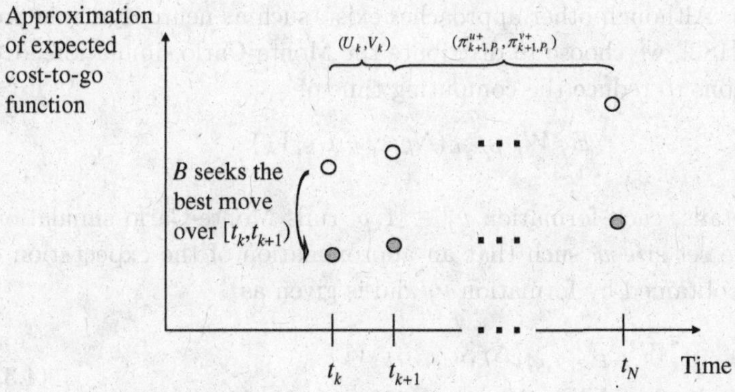

Fig. 4.14 Policy improvement step performed over $[t_k, t_{k+1})$. The best move is defined as the one that minimizes the worst-case cost-to-go function from t_k to t_N.

perfectly known, and $\widetilde{V}_{k+1,P_i,\gamma}$ must be updated as the information on the TC changes. Therefore, we propose to avoid calculating $\widetilde{V}_{k+1,P_i,\gamma}$, and instead approximate the expectation of $W'_{k,P_i,\gamma}$ by means of Monte-Carlo simulations, as explained in the sequel.

Referring to Fig. 4.14, for the entire team of UAS, the policy improvement step consists of selecting and applying U_k over $[t_k, t_{k+1})$ such that $W_{k,P_i,\gamma}$, valued at some to-be-determined TC $S_{\gamma,k}$, is minimized starting from current state $(N_k, c_{c,k})$, at time t_k, until time t_N; that is,

$$\mu_{k,P_i}(N_k, c_{c,k}) = \arg \min_{\substack{U_k \in \mathcal{U}_k(r_{c_{d,k}}) \\ r_{c_{d,k}} \in R_{c_{c,k}}}} \max_{V_k} W_{k,P_i,\gamma}. \qquad (4.32)$$

At this stage, two issues have to be addressed; namely, the computation of the TC estimate, $S_{\gamma^E,k}$, $\gamma^E \in [1, \gamma']$, and the computing time needed to evaluate the expectation of $W'_{k,P_i,\gamma^E}(N_k, c_{c,k}, U_k, V_k)$ for $\gamma^E \in [1, \gamma']$ and all possible pairs (U_k, V_k).

A multi-step lookahead policy may result in better performance than a one-step lookahead policy. The m-step approach requires that the minimization be carried out by seeking a sequence of moves U_j, $j = k, ..., mk-1$ over $[t_k, t_{mk})$, where m is an integer greater than one. However, a gain in performance is obtained at the cost of an increase in computational load.

Approximation of expected cost-to-go function Evaluating W_{k,P_i,γ^E} in Eq. (4.32) necessitates the computation of the expectation of W'_{k,P_i,γ^E} in Eq. (4.31), which can be done by means of Monte-Carlo simulations. Although other approaches exist, such as neuro-dynamic programming [182], we choose to distribute the Monte-Carlo simulations over the formations to reduce the computing time of

$$\underset{N_{k+1}}{E_\gamma}\, W'_{k,P_i,\gamma^E}(N_k, c_{c,k}, U_k, V_k).$$

In more details, each formation $\nu_j \in [1,p]$ runs Monte-Carlo simulations with sample set size η_l such that an approximation of the expectation of W'_{k,P_i,γ^E} is obtained by formation ν_j and is given as

$$\begin{aligned}&\widetilde{W'}_{k,P_i,\gamma^E,\nu_j}(N_k, c_{c,k}, U_k, V_k)\\ &= \tfrac{1}{\eta_l}\textstyle\sum_{i=1}^{\eta_l} W'_{k,P_i,\gamma^E,\nu_j}(N_k, c_{c,k}, U_k, V_k)[i],\end{aligned} \tag{4.33}$$

where $[i]$ denotes the realization of MDPs (4.6)-(4.7) in closed loop with the base policy, at the ith step. The result of the realization is utilized to compute $W'_{k,P_i,\gamma^E}(N_k, c_{c,k}, U_k, V_k)$. Index ν_j is included in Eq. (4.33) to specify that $\widetilde{W'}$ is computed by formation ν_j. Once the $\widetilde{W'}_{k,P_i,\gamma^E,\nu_j}, \nu_j \in [1,p]$, are calculated, a consensus is required so that each formation agrees on a single value, which represents the multi-formations approximation of the expected cost-to-go function. The latter is given by

$$\begin{aligned}&\underset{N_{k+1}}{E_\gamma}\, W'_{k,P_i,\gamma^E}(N_k, c_{c,k}, U_k, V_k)\\ &\simeq \tfrac{1}{p}\textstyle\sum_{j=1}^{p} \widetilde{W'}_{k,P_i,\gamma^E,\nu_j}(N_k, c_{c,k}, U_k, V_k).\end{aligned} \tag{4.34}$$

Consensus on the cost-to-go function can be obtained by sharing $\widetilde{W'}_{k,P_i,\gamma^E,\nu_j}$ among the UAS formations. To do so, the vehicles must have the ability to communicate. The computation of approximation (4.34) is, afterwards, done by each formation. Once consensus is obtained, the roll-out policy $\mu_{k,P_i}(N_k, c_{c,k})$ is obtained from Eq. (4.32), where W_{k,P_i,γ^E} in Eq. (4.31) is replaced by its approximation given as

$$\sum_{\nu\in B} AU^{\nu}_{2,k} + \widetilde{W'}_{k,P_i,\gamma^E,\nu_j}. \tag{4.35}$$

When the number of formations is relatively large, the communication network is possibly connected rather than being complete. Thus, each formation is confined to calculating the average in approximation (4.34) using only local information; that is, information shared with neighbors. To guarantee that formations reach an agreement, a consensus algorithm for multi-agent networked systems must be used. By exploiting local, hence partial, information, the consensus algorithm enables complying to network bandwidth limitations, although at a cost of transients that are inherent to such algorithms. The size η_l of the sample set serves as a tunable parameter that can help comply with real-time constraints; however, a possible deterioration in the quality of the approximation is expected if η_l is too small.

There is a panoply of algorithms enabling consensus reaching for a set of agents. References [183–189] provide detailed information on consensus algorithms.

Selection of TC The TC is selected over $[t_k, t_{k+1})$ by means of the maximum *a posteriori* (MAP) estimate given by

$$\gamma^E = \arg\max_{\gamma \in [1,\gamma']} P(S_{\gamma,k} \mid \mathrm{z}_k), \tag{4.36}$$

where the information state is obtained from the RBF in Eqs. (4.25)-(4.29).

Alternatives to the MAP estimate can be utilized to calculate the TC. The risk-sensitive control approach [164–166, 190], for instance, results in a TC obtained according to

$$\gamma^E = \arg\max_{\gamma \in [1,\gamma']} (P(S_{\gamma,k} \mid z_k) + \kappa \min_{\substack{U_k \in \mathcal{U}_k(r_{c_{d,k}}) \\ r_{c_{d,k}} \in R_{c_{c,k}}}} \max_{V_k} W_{k,P_i,\gamma}(N_k, c_{c,k}, U_k, V_k)).$$

Setting κ to 1 corresponds to the deterministic game certainty equivalence principle [164]. Such an approach requires, however, the online computation of $W_{k,P_i,\gamma}$, for all $\gamma \in [1, \gamma']$, which may lead to a prohibitive computational load depending on the value of η_l. The suboptimal MAP estimate in Eq. (4.36) is thus preferred to keep the blue-team decision policy tractable in a computational sense. It should be noted that the performance of the closed-loop system depicted in Fig. 4.11 depends on the rate of convergence of the filter as compared with the time constant associated with variations of the TC, if any. This is true with any estimator-based control scheme. A filter with a slow time constant, when compared with variations in the TC, may result in degraded closed-loop performance.

4.1.8.3 *Computational load*

Blue-team formations share their states N_k^ν and X_k^ν via a communication network, as mentioned in Assumption 4.2. Transmission of information among UAS and SVs, and computations onboard the vehicles should comply with the schedule shown in Fig. 4.8. The schedule is shown in more details in Fig. 4.15. The figure shows that the information state $P(S_{\gamma,k} \mid \mathrm{z}_k)$ is

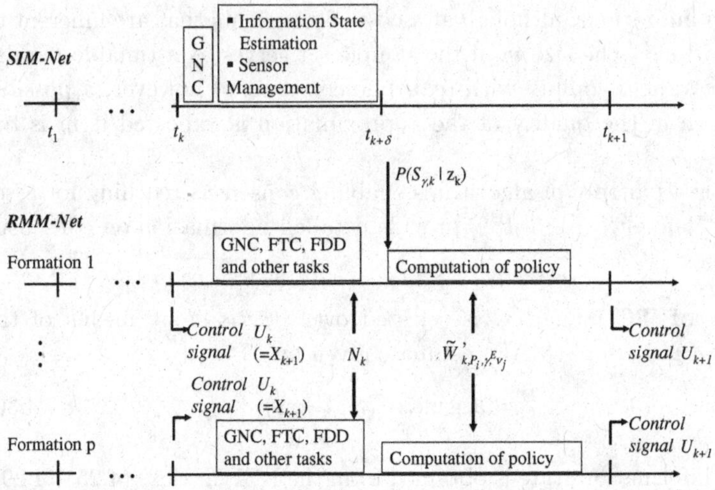

Fig. 4.15 Scheduling of tasks involved in computing the decision policy.

obtained by the UAS from data communicated from SIM-Net to RMM-Net. Furthermore, state variable N_k of all the formations and the approximations on the expected cost-to-go functions $\widetilde{W'}_{k,P_i,\gamma^E,\nu_j}$ are shared among the UAS via RMM-Net. Such communications are needed for a decentralized implementation of the policy.

Computing the base policy is more demanding than obtaining the rollout policy. Hence, the base policy used by the UAS to plan their paths from a given target area T_k to target area T_{k+1} should not be re-calculated online to adapt to changes in the estimates of the TC when the UAS are flying from T_k to T_{k+1}. Instead, a heuristic approach should be used online. However, when the UAS are located in the portion of the theater between target T_k and target T_{k+1}, a distributed computation of the base policy can be done to plan the paths from T_{k+1} to T_{k+2}, from T_{k+2} to T_{k+3}, and so on. Such approach allows using the latest information on the TC, provided the available resources can handle the increase in computational and communication loads. To understand the number of computations involved, consider the value function in Eq. (4.16), which depends on two states; namely, $N_j \in \{0, 1, ..., n_p\}^p$ whose cardinality is $(n_p + 1)^p$ and $c_{c,j}$, which represents one of the $Cb(p_j, m_j)$ possible distributions of p_j formations among the m_j nodes of E_j. For each $(N_j, c_{c,j})$ pair, the best policy is sought in $U_j(r_{c_{d,j}})$, where

$$r_{c_{d,j}}(c_{c,j}) = c_{d,j+1} \in E_j$$

represents the authorized transition from $c_{c,j}$ to $c_{c,j+1}$. Note that the total number of authorized transitions from E_j to E_{j+1} is $Cb(p_j, b_j)$, where b_j is the number of edges that relate E_j to E_{j+1}, with $j \in \{1, ..., N_N\}$, and

$$N_N = \frac{d_m}{e} + i + 1.$$

The distributed computation of (4.34) is performed online from a given pair $(N_j, c_{c,j})$ and requires $O(R_j p \eta_l (N_N - j))$ operations, where

$$R_j << Cb(p_j + b_j - 1, b_j - 1)$$

is the number of transitions from $c_{c,j}$. The computational load mainly depends on η_l used in Eq. (4.33).

4.1.9 *Design of CHM and decision making system*

We give a brief summary of the key steps in the design of the CHM and decision making system.

4.1.9.1 *Perfect information*

Problem 4.1 is solved as follows.

Step 1. Obtain the actual TC $S_{\gamma^+,1}$ for the given grid G_{UT}.

Step 2. Set initial values for p, n_p, and m_p.

Step 3. Determine locations of the base B_a and the target areas T_i. The order of the visits is given by a formation-to-target assignment algorithm, such as that proposed in Ref. [155]. The sequence of targets is assumed known by the cooperating UAS prior to mission.

Step 4. Obtain the optimal planning and maneuvering decision policy by solving Eq. (4.16) following Proposition 4.1. If more than one sequence of paths and maneuvers minimizes Eq. (4.13), randomly select among those sequences the policy that will be actually implemented.

Step 5. Store the decision policy as a lookup table. During mission, at each decision time instant t_k, a UAS formation determines at which node it should be located at the next decision point, t_{k+1}, from the knowledge of $S_{\gamma^+,1}$, N_k^ν, and X_k^ν, for all ν, i and j, by simply selecting the corresponding entry in the lookup table.

4.1.9.2 *Partially known environment*

Problem 4.2 is solved as follows.

Step 1. Offline, calculate the base policy as is done in steps 1-4 of the algorithm that solves Problem 4.1. The TC selected for the offline computations, $S_{\gamma^+,1}$, corresponds to the most harmful configuration. Alternatively, one may estimate the initial TC using Eq. (4.30) or else.

Step 2. Design an online state estimator with Eq. (4.25). Solve the sensor allocation problem, with Eq. (4.28) or Eq. (4.29), depending on the availability of measurements.

Step 3. The one-step lookahead policy improvement step is obtained as follows. Over each interval $[t_k, t_{k+1})$, the approximation on the expected cost-to-go functions $\widetilde{W'}_{k,P_i,\gamma^E,\nu_j}$ is obtained by each formation by carrying out Monte-Carlo simulations. Then, $\widetilde{W'}_{k,P_i,\gamma^E,\nu_j}$ are shared among the UAS via RMM-Net to reach a consensus. The policy is obtained by solving Eq. (4.32), where $W_{k,P_i,\gamma}$ is replaced by expression (4.35). Decision variable is applied over the next time interval $[t_{k+1}, t_{k+2})$. The block diagram of the overall system is given in Fig. 4.11.

4.2 Cooperation Despite Information Flow Faults

4.2.1 *Context*

To reach the target areas with a maximum number of operational, or healthy, vehicles and maneuvering capabilities, teaming UAS rely heavily on communications, as discussed in Sec. 4.1. The health state must be shared among the formations through the wireless communication network to obtain optimal, or near-optimal, path planning and maneuvering. Health state refers to current levels of maneuvering capability, or energy, of the UAS formations. No matter how reliable is the network, it is subject to failures, delays and information loss due to a variety of factors [24]. Failure to transmit and/or receive health state information through RMM-Net is expected to significantly reduce the effectiveness of the decision making system. Let us assume that the second network, denoted as SIM-Net, is available at times when RMM-Net is down. Such assumption in fact enables a certain level of redundancy in the communications. We thus assume that the probability of information loss for the communications between the SVs and the UAS is likely to be smaller than that characterizing information loss in the communications among the UAS. In this context, we propose to leverage the availability of SIM-Net to calculate a probabilistic distribution on the possible number of healthy UAS in each formation, and to design a RBF-based health state estimator which feeds the RMM computing nodes, in each formation, in case of intermittent losses of communications.

Consider Fig. 4.16. In part (a), formations 1, 2 and p are shown at a time instant between t_k and t_{k+1}. The formations share information through RMM-Net (dashed lines) and SIM-Net (dotted lines). Communications among the UAS pertain to their state (health and position) and the approximation on the expected cost-to-go function. The latter is denoted as intermediate data in the figure. The complete state information is available to the formations between t_k and t_{k+1}. A communication breakdown occurs for a certain period of time between t_{k+1} and t_{k+2}. This situation is illustrated in Fig. 4.16(b). According to part (b) of the figure, the communication links between formations 1 and p, and between formations 2 and p are broken. Formation p no longer sends information to formation 2. Formation 1 has no transmission to formation p. Irrespective of the source of the information flow faults, the formations have a reduced capability to share information, and hence subpar performance of the decision policies is expected. Information flow fault may arise due to Tx/Rx component failures, presence of large buildings, or long inter-formation distances, to cite a few. For the situation illustrated in Fig. 4.16(b), the CHM and decision making system of formation 2 either relies on past information or calculates an estimate of the state and intermediate data associated with formation p. The cooperative system of formation p proceeds similarly for the missing information on formation 1.

4.2.2 *Impact of information flow fault on UAS decision policies*

Section 4.1 presents baseline and one-step lookahead rollout policies that provide near-optimal routing and maneuvering management of UAS formations despite a risky environment and partial knowledge of TC. UAS-threat encounters are formulated as a stochastic game. Figures 4.8 and 4.15 show the communication and computing tasks required to carry out the decision making. From those figures and the example of Fig. 4.16, it becomes clear that the loss of information due to RMM-Net communication breakdowns will impede sharing N_k^{ν} and $\widetilde{W'}_{k,P_i,\gamma^E,\nu_j}$ for some or all formations $\nu \in B$, depending on the resulting (faulty) communication graph topology. Note

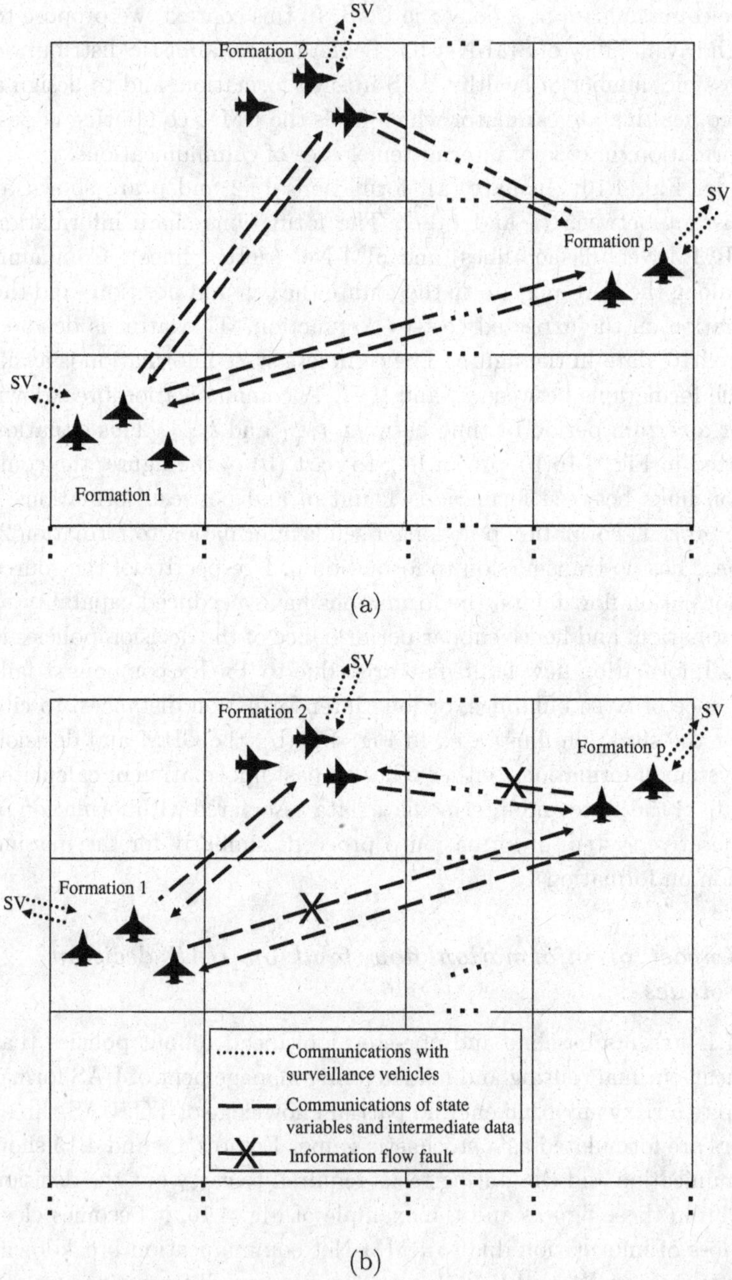

Fig. 4.16 Example of information flow fault.

that $c_{c,k}$ in $\widetilde{W}'$ (4.33), which represents the position of the formations at t_k, is known from each formation at t_k since $X_k = U_{1,k-1}$ is computed by each formation. The timely knowledge of N_k^ν is critical since the loss of a UAS can occur at any time over $[t_k, t_{k+1})$, while the health state of the formations must be known by each formation at $t_{k+\delta}$. At time $t_{k+\delta}$, the computations for the selection of the next edge of the path are in progress, as shown in Fig. 4.17 for the case of a fault in the information supplied by formation p to formation 2. This situation is also illustrated in Fig. 4.16(b). UAS formation 2 requires extra processing to carry out the health state estimation process and the approximation of the cost-to-go function. In general, when the information on the health state of the formations is unavailable, a system either relies on past, possibly obsolete, information, or on an estimate of N_k^ν, for some or all formations $\nu \in B$. To carry out the estimation process, we want to leverage the ability of the SVs to detect and classify threat types, as presented in Sec. 4.1, by designing an additional RBF, labeled as the N_k-information state RBF. For this purpose, consider the following assumption on the information flow faults.

Assumption 4.3. Availability of Information. At any time t_k, N_{k-1} and V_{k-1} are assumed available to SVs prior to an RMM-Net breakdown. RMM-Net information flow fault occurs intermittently or permanently over interval $[t_k, t_{k+\delta}]$, where $t_{k+\delta} < t_{k+1}$. Furthermore, U_{k-1} is available to SVs from the measurements. SVs rely on a graph of feasible paths and have knowledge of the grid G_{UT}. An information flow fault is defined as the loss of the health state and expected cost-to-go function of one or more formations over a single interval $[t_k, t_{k+1}]$ in RMM-Net, as shown in the example of Fig. 4.17.

4.2.3 *Health state estimation*

During operation, cooperating UAS and SVs estimate the threat configuration by means of the TC-information state RBF given by Eqs. (4.24)-(4.29). A second filter, labeled as the N_k-information state RBF, yields the estimate on the N_k-information state. This filtering process is performed on the SIM-Net computing nodes. The output of the filter is transmitted to each UAS formation part of RMM-Net. The N_k-information state RBF depends on the TC-information state estimate at t_k, and on the last available health state, assumed to be N_{k-1}. SIM-Net estimates the TC with RBF (4.25). Then, exploiting available information, as stated in Assump-

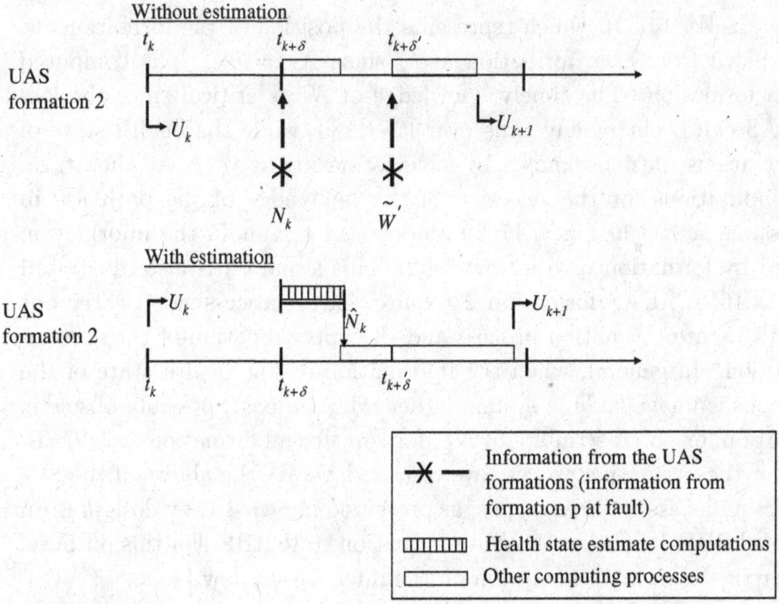

Fig. 4.17 Information flow fault and impact on computations.

tion 4.3, SIM-Net provides an estimate on the health state N_k of the UAS formations. As shown in Fig. 4.18, two cooperating UAS formations are faced with RMM-Net communication breakdowns (dotted lines) while the network of SVs is not encountering information flow faults (solid lines). SVs can thus share information to provide estimates on the TC and on the health state of the formations, and are assumed capable of communicating such information to the UAS.

Assume that RMM-Net is down over $[t_k, t_{k+\delta}]$. An estimate of N_k is obtained by SIM-Net, which observes the UAS formations, updates the RBF, and sends a distribution on N_k to the formations, by virtue of Assumptions 1 and 2. For the SVs, the measurement model used to estimate the number of healthy UAS of any formation ν_i at $t_{k+\delta}$ is defined in a similar manner to the process in Eq. (4.2). The health state of formation ν_i available at $t_{k+\delta}$ is $N_{k-1}^{\nu_i}$, which is assumed known. Here, one aims to estimate $N_k^{\nu_i}$. At $t_{k+\delta}$, $N_k^{\nu_i} \in \{0, 1, 2, ..., N_{k-1}^{\nu_i}\}$ and measurement $\zeta_k^{sv} \in \{N_k^{\nu_i}, nd\}$. A value of nd indicates that the detection process has failed. Index sv refers to the SV observing formation ν_i engaged along (i, j). The measurement model

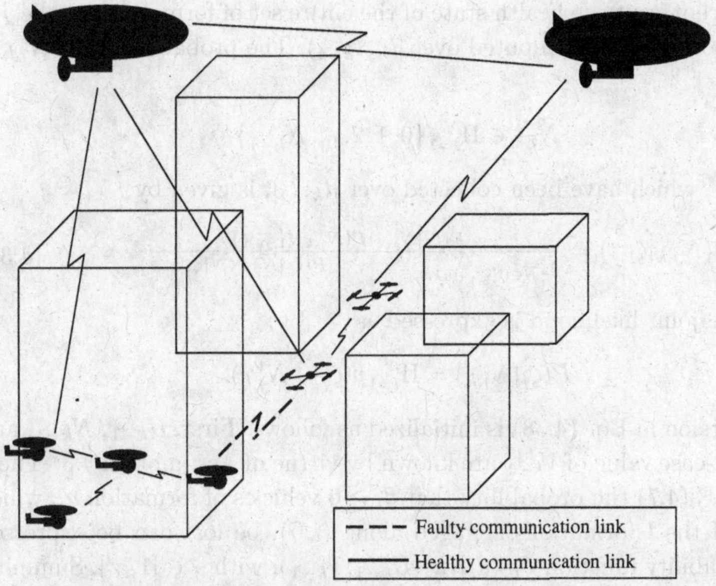

Fig. 4.18 High-altitude SVs transmitting information to low-altitude UAS formations subject to information flow fault.

of ν_i is thus

$$p(\zeta_k^{\nu_i} \mid N_k^{\nu_i}) = \begin{cases} p_d^{sv}(\delta_{ij,k}) L(\zeta_k^{\nu_i} \mid N_k^{\nu_i}) & \text{if} \quad \zeta_k^{\nu_i} \neq nd, \\ 1 - p_d^{sv}(\delta_{ij,k}) & \text{if} \quad \zeta_k^{\nu_i} = nd, \end{cases} \tag{4.37}$$

$$L(\zeta_k^{\nu_i} \mid N_k^{\nu_i}) = \begin{cases} p_c^{sv}(\delta_{ij,k}) & \text{if} \quad \zeta_k^{\nu_i} = N_k^{\nu_i}, \\ \frac{1-p_c^{sv}(\delta_{ij,k})}{N_{k-1}^{\nu_i}} & \text{if} \quad \zeta_k^{\nu_i} \neq N_k^{\nu_i}, \end{cases}$$

where p_d^{sv} and p_c^{sv} denote the probability of detection and classification of formation ν_i by sv. Equation (4.37) is reminiscent of Eq. (4.2). By classification, it is meant the process by which a value is assigned to the observation variable $\zeta_k^{\nu_i}$ given the actual health state of ν_i being $N_k^{\nu_i}$. Observation variable $\zeta_k^{\nu_i}$ can take values in $\{0, 1, 2, ..., N_{k-1}^{\nu_i}\}$. In (4.37), $\delta_{ij,k}$ is the distance that separates ν_i, engaged along (i, j), from sv. The following notation is adopted in the sequel:

$$\zeta_k = \{\zeta_k^1, ..., \zeta_k^{\nu_p}\},$$

$$\zeta_{1,k} = \{\zeta_1, ..., \zeta_k\}.$$

Let $N_{j,k}$ represent the health state of the entire set of formations at the jth iteration of the filter computed over $[t_k, t_{k+\delta})$. The probability that $N_{j,k}$ is equal to

$$\widetilde{N}_{j,k} \in \Pi_{i=1}^{p}\{0, 1, 2, ..., N_{k-1}^{\nu_i}\}$$

given $\zeta_{1,j}$, which have been collected over $[t_k, t_j]$, is given by

$$P(\widetilde{N}_{j,k}|\zeta_{1,j}) = \frac{P(\zeta_j|\widetilde{N}_{j,k})P(\widetilde{N}_{j,k}|\zeta_{1,j-1})}{\sum_{\overline{N}\in\Pi_{i=1}^{p}\{0,1,2,\ldots,N_{k-1}^{\nu_i}\}} P(\zeta_j|\overline{N})P(\overline{N}|\zeta_{1,j-1})}, \tag{4.38}$$

where the joint likelihood is expressed as

$$P(\zeta_j|\widetilde{N}_{j,k}) = \Pi_{i=1}^{p} p(\zeta_j^{\nu_i} \mid N_{j,k}^{\nu_i}).$$

The recursion in Eq. (4.38) is initialized as follows. First, U_{k-1}, N_{k-1}, and the worst-case value of V_{k-1} are known by virtue of Assumption 4.3. Then, from MDP (4.7) the probability that $d_i \geq 0$ vehicles of formation ν_i, which is part of the l formations engaged along (i, j), be lost can be expressed as a probability function $P_{\nu_i}(d_i, l, \gamma, U_{k-1}, V_{k-1})$, with $\gamma \in [1, \gamma']$. Summing with respect to γ yields from Eq. (4.25) the marginal distribution

$$\begin{aligned} &P_{\nu_i}(d_i, l, U_{k-1}, V_{k-1} \mid \mathrm{z}_k) \\ &= \textstyle\sum_{\gamma\in[1,\gamma']} P(S_{\gamma,k} \mid \mathrm{z}_k) P_{\nu_i}(d_i, l, \gamma, U_{k-1}, V_{k-1}), \end{aligned} \tag{4.39}$$

which is selected to initialize Eq. (4.38) as follows

$$\begin{aligned} P(\widetilde{N}_{1,k}|\zeta_{1,1}) &= [P(\widetilde{N}_{1,k}^{\nu_1}), ..., (\widetilde{N}_{1,k}^{\nu_p})], \\ P(\widetilde{N}_{1,k}^{\nu_i}) &= P_{\nu_i}(d_i, l, U_{k-1}, V_{k-1} \mid \mathrm{z}_k), \end{aligned} \tag{4.40}$$

where

$$\widetilde{N}_{1,k}^{\nu_i} = N_{k-1}^{\nu_i} - d_i$$

for $d_i \in \{0, 1, ..., N_{k-1}^{\nu_i}\}$.

Once $P(\widetilde{N}_{j,k}|\zeta_{1,j})$ is obtained, from Eq. (4.38), a MAP estimate $\widehat{N}_{k,j}$

$$\widehat{N}_{k,j} = \underset{\widetilde{N}_{j,k}\in\Pi_{i=1}^{p}\{0,1,2,\ldots,N_{k-1}^{\nu_i}\}}{\arg\max} P(\widetilde{N}_{j,k}|\zeta_{1,j})$$

is utilized in Eq. (4.32) and approximation (4.35) by fixing $\widehat{N}_k$ to $\widehat{N}_{k,j^o}$. The j^oth step of the RBF in Eq. (4.38) is defined such that the computation of the rollout policy takes place during time interval $[t_{j^o}, t_{k+1})$. The MAP estimate is attractive for implementation purposes, owing to its limited computational load. As already mentioned for the TC-information state

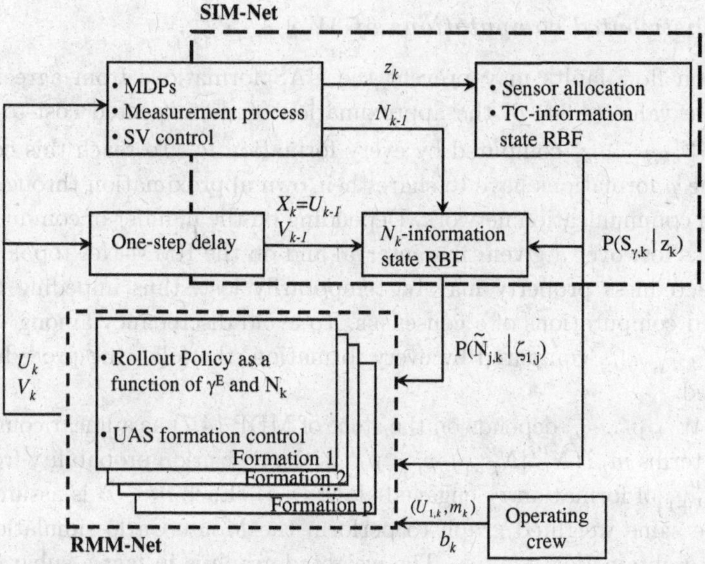

Fig. 4.19 Block diagram of CHM and decision making system in closed loop with health state and TC estimators.

RBF, more advanced, although still suboptimal, approaches exist. It is worth mentioning that we suppose RMM-Net information flow faults occur intermittently or permanently over interval $[t_k, t_{k+\delta}]$, where $t_{k+\delta} < t_{k+1}$, from Assumption 4.3. However, to carry out the N_k-state information estimate and the calculation of the policy, there is an upper bound on the acceptable time $t_{k+\delta}$. An alternative to the N_k-state information estimate is to use the available health state at t_{k-1}, N_{k-1} [123].

Closed-loop dynamics composed of MDPs (4.32)-(4.36) and (4.24)-(4.29) are now augmented with the N_k-information state RBF given by Eqs. (4.38)-(4.39). Figure 4.19 shows the block diagram of the CHM and decision making system in closed loop with the health state estimator labeled as N_k-information state RBF, and with the TC-information state RBF. The sensor allocation and the TC-information state RBF are instrumental to provide a MAP TC estimate. This estimate is used to compute the rollout policy and the N_k-information state RBF, if need be, as shown in the figure. The N_k-information state filter leverages available information about the formations to provide a MAP estimate of the missing health state required to compute the rollout policy.

4.2.4 *Distributed computations of* $\widetilde{W'}_{k,P_i,\gamma^E,\nu_j}$

Information flow faults may prevent the UAS formations from agreeing on a single value (4.34) of the approximation of the expected cost-to-go function $\widetilde{W'}_{k,P_i,\gamma^E,\nu_j}$ computed by every formation ν_j. To reach this consensus, the p formations have to share their own approximation through a connected communication network. Depending on the number of communication links lost over a given time interval and on the RMM-Net topology, the connectedness property may be temporarily lost, thus impeding the distributed computations of a consensus. To avoid discrepancy among values of $\widetilde{W'}_{k,P_i,\gamma^E,\nu_j}$ computed by every formation, the following procedure is proposed.

First, $\widetilde{W'}_{k,P_i,\gamma^E,\nu_j}$ depends on the state of MDP (4.7) as a linear combination of terms $m_p I(N_k^{\nu_j}|N_{k-1}^{\nu_j}) \cdot p_{i,j}^{\nu_j}$. $p_{i,j}^{\nu_j}$ is the transition probability from $N_k^{\nu_j}$ to $N_{k+1}^{\nu_j}$ of formation ν_j engaged along (i,j). Each $\nu_j \in B$ is assumed to use the same weighted graph to perform the Monte-Carlo simulations required to obtain $\widetilde{W'}_{k,P_i,\gamma^E,\nu_j}$. The weighted graph is in fact a subgraph of $\mathcal{G}(P_i)$ with weights corresponding to MDP (4.7). At t_k, the subgraph starts from the actual distribution of the formations, which is one element of $C(P_i, B, E_k)$ in Eq. (4.3), and ends at the node where stands the target. For example, consider the subgraph shown in Fig. 4.20. The example first described in Fig. 4.7 is revisited. At t_4, the formations are located at nodes 7, 13, and 19, which corresponds to triplet $(7,13,19) \in C(P_i, B, E_4)$. The target area is at node 22. The weights found on the edges are conditional probabilities of MDPs. When formation ν_k is set to engage along edge (i,j), $p_{i,j}^{\nu_k} = p_{i,j}$.

Then, an algorithm is implemented in each ν_j such that the set of probabilities $\{p_{i,j}^{\nu_j}, \nu_j \in B\}$ yields at t_k and at a given iteration of the Monte-Carlo simulation, an identically realized state transition $N_k \to N_{k+1}$. The algorithm is composed of a uniform pseudorandom number generator (PRNG), such as the Mersenne twister [191], combined with the inverse transform sampling [192]. The internal state s_o of PRNG must be the same in every $\nu_j \in B$ so that each formation computes the same state transition, $N_k \to N_{k+1}$. Suppose for example that, at t_k, and at the ith iteration of the Monte-Carlo simulation, PRNG(s_o) provides in each formation the pseudorandom number $s_k[i] \in [0,1]$. Furthermore, assume that the maximum number of UAS loss in formation ν_j after an encounter with a threat is given by d^{ν_j}. Let

$$p_{x,i,j}^{\nu_j} = P(N_{k+1}^{\nu_j} = N_k^{\nu_j} - x),$$

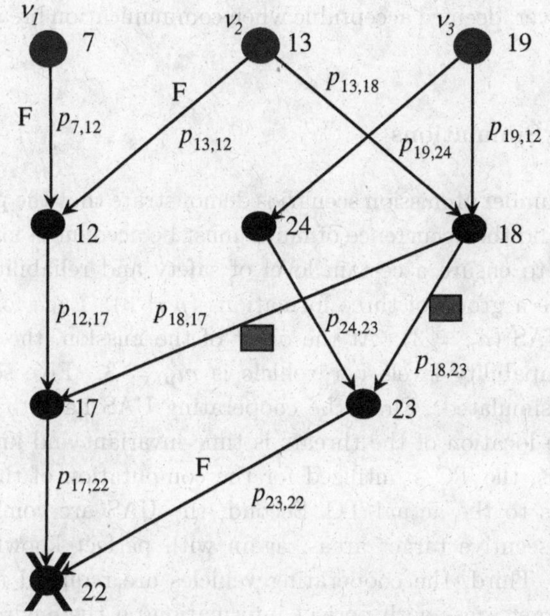

Fig. 4.20 Possible edges for formations ν_1, ν_2, and ν_3 from nodes (7,13,19) to node 22 (target denoted as X).

where x is an integer in $[0, d^{\nu_j}]$. Then, $s_k[i]$ is compared to values that $\{p^{\nu_j}_{x,i,j}, \nu_j \in B\}$ can take; that is, every formation computes the state

$$N_{k+1} = \{N^1_{k+1}, ..., N^{\nu_j}_{k+1}, ..., N^p_{k+1}\}$$

such that

$$N^{\nu_j}_{k+1} = \begin{cases} N^{\nu_j}_k & \text{if} \quad s_k[i] \in [0, p^{\nu_j}_{0,i,j}), \\ N^{\nu_j}_k - 1 & \text{if} \quad s_k[i] \in [p^{\nu_j}_{0,i,j}, p^{\nu_j}_{0,i,j} + p^{\nu_j}_{1,i,j}), \\ \vdots & \\ N^{\nu_j}_k - d^{\nu_j} & \text{if} \quad s_k[i] \in [1 - p^{\nu_j}_{d^{\nu_j},i,j}, 1]. \end{cases} \tag{4.41}$$

By applying the above procedure when RMM-Net is operational, Monte-Carlo simulations are systematically carried out on a limited set of realizations of state transitions $N_k \rightarrow N_{k+1}$ obtained with a common seed. However, with consensus, a richer and a larger sample set is used by the system; thus, resulting in a smaller variance between the expected cost-to-go function and its approximation. If RMM-Net is operational, the UAS should therefore proceed with a consensus algorithm. The above procedure is, however, deemed acceptable when communication breakdowns are intermittent.

4.3 Numerical Simulations

Simulations of a number of mission scenarios demonstrate that the presence of health threats and the occurrence of faults must be accounted for by the cooperating UAS to ensure a certain level of safety and reliability. The simulations involve a group of three formations ($p = 3$). Each formation comprises three UAS ($n_p = 3$). At the onset of the mission, the number of maneuvering capability levels per vehicle is $m_p = 3$. Five scenarios are modeled and simulated. First, the cooperating UAS have to reach a single target. The location of the threats is time invariant and known by B. In other words, the TC s_γ utilized for the computation of the based policy corresponds to the actual TC. Second, the UAS are commanded to reach four consecutive target areas, again with perfect knowledge of the environment. Third, the cooperating vehicles are required to reach four successive target areas with perfect information on the environment, although the MDPs are perturbed to a certain extent. Fourth, the UAS have to reach three successive target areas with a partial knowledge of the TC. Finally, intermittent information flow faults in the communications among the UAS are simulated.

4.3.1 *Single target area and perfectly known environment*

DPE (4.16) is solved with path length constraint $d(B_a, T) = d_m = 6$. Base Ba and target T are located at nodes 24 and 2, respectively, on the grid shown in Fig. 4.21. Fifteen threats have been placed randomly over the grid. As stated earlier, the control law obtained at each discrete-time instant may not be unique. If there are several control signals available for any given N_k and $c_{c,k}$, from the look-up table, the one actually applied to the UAS is selected randomly among the available, equiprobable control signals.

4.3.1.1 *Low-level health threat*

The low-level threat is characterized by

$$P^{(1)}_{b,VehNb} = \begin{bmatrix} 1 \; 0 \; 0 \end{bmatrix}, \tag{4.42}$$

and

$$P^{(2)}_{b,VehNb} = \begin{bmatrix} 1 \; \mathbf{0}_{1\times5} \end{bmatrix}, \; P^{(3)}_{b,VehNb} = \begin{bmatrix} 1 \; \mathbf{0}_{1\times8} \end{bmatrix}. \tag{4.43}$$

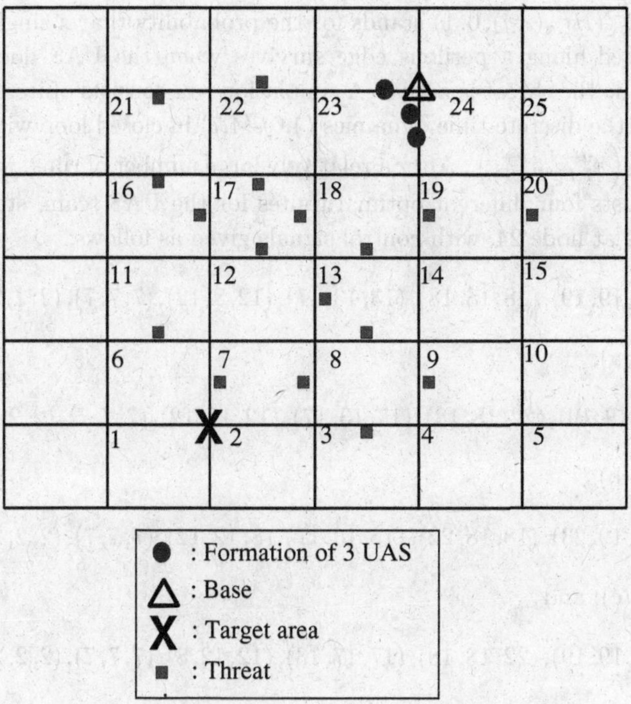

Fig. 4.21 Environment with single target.

Probabilities (4.42) and (4.43) indicate that a maximum of one UAS may be lost during an encounter with a threat regardless of the number of formations engaged along the perilous edge. The scenario relies on the following transition probabilities:

$$p_b^{(1)}(H_{1,k}(i,j), 0, 1) = 0.75,$$

$$p_b^{(1)}(H_{1,k}(i,j), 1, 1) = 0.85,$$

$$p_b^{(2)}(H_{1,k}(i,j), 0, 1) = 0.3,$$

$$p_b^{(2)}(H_{1,k}(i,j), 1, 1) = 0.4,$$

$$p_b^{(3)}(H_{1,k}(i,j), 0, 1) = 0.1,$$

$$p_b^{(3)}(H_{1,k}(i,j), 1, 1) = 0.2.$$

Recall that $p_b^{(1)}(H_{1,k}(i,j),0,1)$ stands for the probability that a single formation engaged along a perilous edge survives when the UAS does not maneuver while the threat is active. A simulation run consists of iterations over $[0,t_6)$ of the discrete-time dynamics (4.6)-(4.7) in closed loop with optimal policies $\left(\pi_{1,P_i}^{u+},\pi_{1,P_i}^{\nu+}\right)$. After a relatively large number of runs, we find that there exists four different optimal routes for the UAS team, starting from the base at node 24, with control signals given as follows:

$$U_1 = \{(19,19,19),(18,18,18),(13,13,17),(12,8,12),(7,7,7),(2,2,2)\}$$

for Fig. 4.22 (a);

$$U_1 = \{(23,19,19),(22,18,18),(17,13,17),(12,12,12),(7,7,7),(2,2,2)\}$$

for Fig. 4.22 (b);

$$U_1 = \{(19,19,23),(18,18,22),(13,13,17),(8,12,12),(7,7,7),(2,2,2)\}$$

for Fig. 4.22 (c); and

$$U_1 = \{(23,19,19),(22,18,18),(17,17,13),(12,12,8),(7,7,7),(2,2,2)\}$$

for Fig. 4.22 (d). For this example, $U_{2,k}^{\nu} = 0$ for all k and for all formations $\nu \in \{1,2,3\}$. This means that the UAS do not actually perform additional maneuvers to reduce their risk of loss. All the threats are active in this example. To evaluate the effectiveness of the proposed decision policy in terms of the total number of maneuvers, or levels of energy, remaining once at close range to target T, that is at time $t = t_6$, five hundred runs are carried out with the aforementioned transition probabilities. For each run, the discrete-time dynamics in closed loop with the policy are simulated. Realization of the probabilities associated with the transition matrices and the randomized control signals are varied from one run to another. The average number of maneuvers remaining at T, denoted as $\widehat{J}$, is computed for two batches of five hundred simulations each. One batch pertains to the proposed multi-formations decision policies. The other batch of five hundred simulation runs illustrates the case where the blue team follows a single route with the following characteristics: the UAS policy is such that $U_{2,k} = 0$, for all k, and the UAS control signals are given as

$$U_1 = \{(19,19,19),(18,18,18),(13,13,13),(8,8,8),(7,7,7),(2,2,2)\}.$$

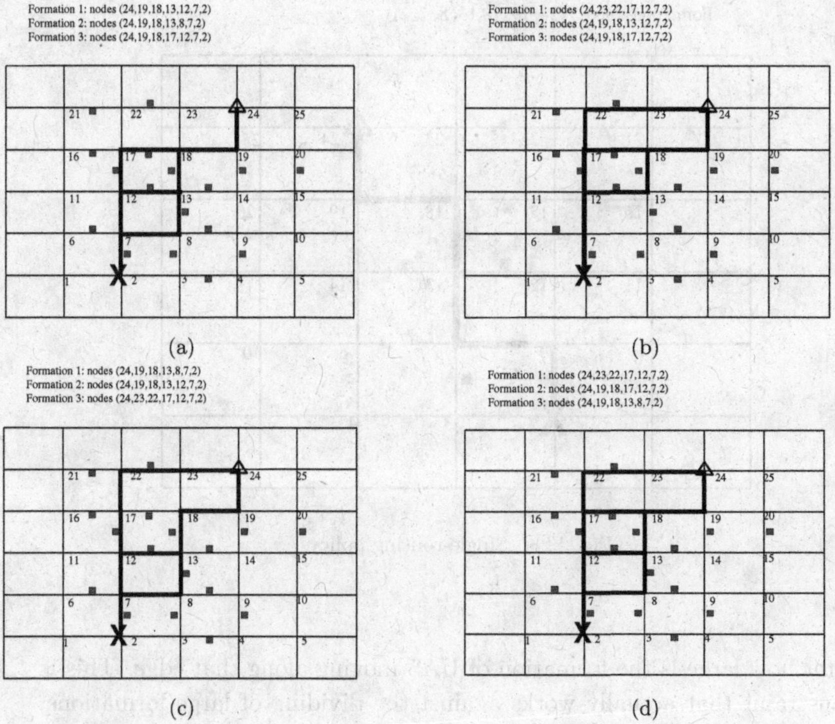

Fig. 4.22 Path Planning for a group of three formations. The proposed decision policy obtained with $d(B_a, T) = 6$ provides four different optimal routes for the multi-formations.

The single path is shown in Fig. 4.23. With the proposed decision policy, $\widehat{J} = 19.6$, whereas with the single-group routing, $\widehat{J} = 18.8$. Recall that the team of UAS initially has available 27 maneuvering capability levels. These results show that the concept of multi-formations provides, on average, an improvement in performance. The simulations confirm that smaller formations are more likely to survive encounters with health-threatening situations than teams of UAS flying as one large group. The simulation results are consistent with the transition probabilities selected for the example, where

$$p_b^{(1)} > p_b^{(2)} > p_b^{(3)} .$$

It should be noted that, in cases of an encounter with a low-level threat, a UAS formation cannot lose more than one UAS along the perilous edge, no

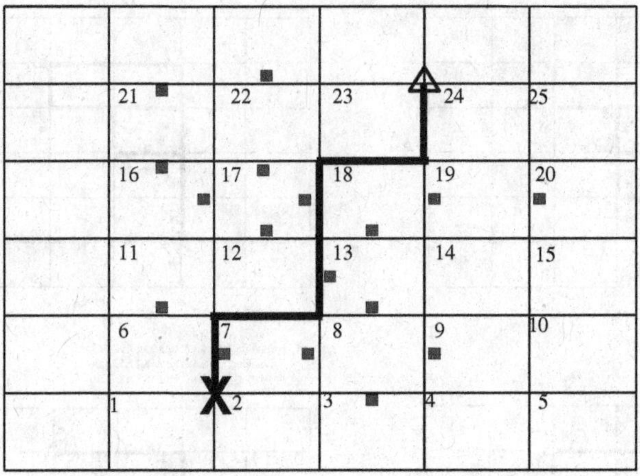

Fig. 4.23 Single-routing policy.

matter how large is the formation of UAS moving along that edge. This is a constraint that actually works against the dividing of large formations, partly explaining why, in this case, the gain in performance is small.

4.3.1.2 *High-level health threat*

Consider encounters with high-level threats resulting in the loss of up to three vehicles. The following distributions $P^{(l)}_{b,VehNb}$ with $l \in \{1, 2, 3\}$ are adopted:

$$P^{(1)}_{b,VehNb} = [1\ 0\ 0], P^{(2)}_{b,VehNb} = [0.85\ 0.15\ \mathbf{0}_{1\times4}], \qquad (4.44)$$

$$P^{(3)}_{b,VehNb} = [0.5\ 0.35\ 0.15\ \mathbf{0}_{1\times6}]. \qquad (4.45)$$

These distributions express the fact that groups of formations are prone to losing several vehicles. The average number of maneuvers remaining at target T after five hundred simulation runs with the proposed decision policy is $\widehat{J} = 17.3$ with a standard deviation $\widehat{\sigma} = 4.1$. The single-group routing results in $\widehat{J} = 13.6$ and $\widehat{\sigma} = 4.4$. A gain of 3.7 maneuvering capability levels, or 13.7% of the number of maneuvering capability levels at base B_a,

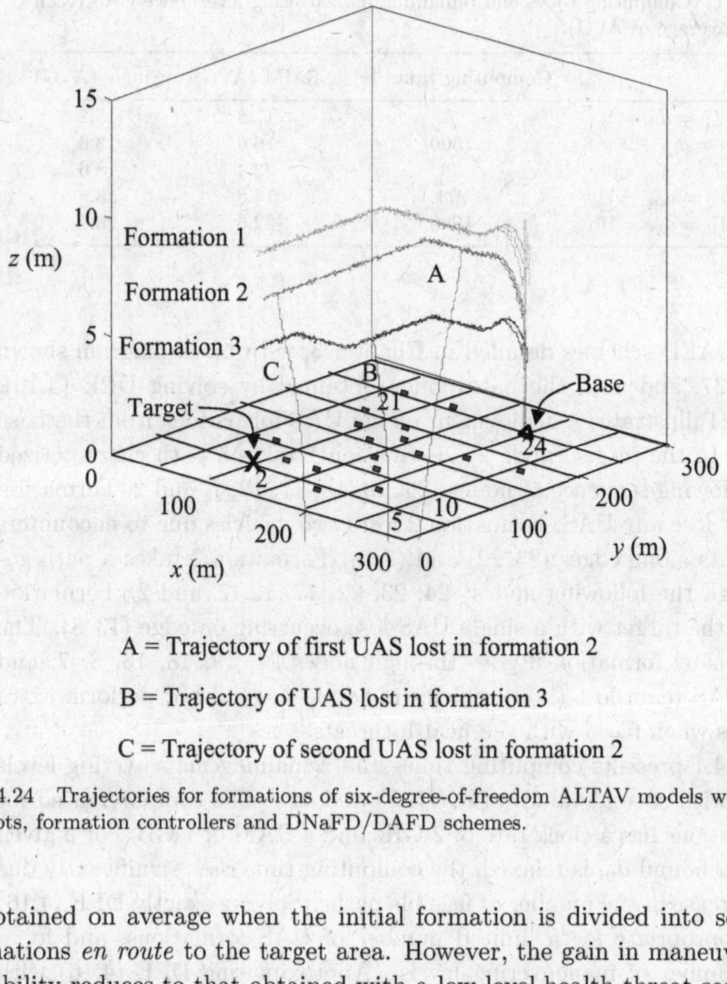

Fig. 4.24 Trajectories for formations of six-degree-of-freedom ALTAV models with autopilots, formation controllers and DNaFD/DAFD schemes.

is obtained on average when the initial formation is divided into several formations *en route* to the target area. However, the gain in maneuvering capability reduces to that obtained with a low-level health threat scenario when $P^{(l)}_{b,VehNb}$, $l \in \{1,2,3\}$, approaches elementwise $[1 \ \mathbf{0}_{1\times n_p(l-1)}]$. For instance, when

$$P^{(3)}_{b,VehNb} = [0.65 \ 0.25 \ 0.1 \ \mathbf{0}_{1\times 6}],$$

the proposed control law yields $\widehat{J} = 17.8$, whereas the single-group routing results in $\widehat{J} = 15.2$. Figure 4.24 presents trajectories for the three formations of UAS. Each formation comprises three UAS flying in a string-type geometry. The leader vehicle is at the front of the group. The trajectories are those of the six-degree-of-freedom ALTAV mathematical models in closed loop with the low-level controllers, formation controllers and

Table 4.1 Computing times and remaining maneuvering levels once UAS reach target (average or AVG).

	Computing times (s)	RMM (AVG)	Single (AVG)
$d(B_a, T) = d_m = 6$	7	17.8	15.2
$d(B_a, T) = d_m + 2 = 8$	1500	16.6	13.3
$d(B_a, T) = d_m = 9$	36	22.4	18.6
$d(B_a, T) = d_m = 11$	271	14.8	13.5
$d(B_a, T) = d_m = 16$	1980	22.6	20.7

DNaFD/DAFD schemes detailed in Chapter 3, with block diagram shown in Fig. 3.27, and with the base policy obtained by solving DPE (4.16). Figure 4.24 illustrates a deployment of the UAS formations from the base (node 24) to the target (node 2). Formation 1 takes a path characterized by the following sequence of nodes: 24, 19, 18, 17, 12, 7, and 2. Formation 1 does not lose any UAS. Formation 2 loses two vehicles due to encounters with threats along edges $(23, 22)$ and $(7, 2)$. Formation 2 takes a path going through the following nodes: 24, 23, 22, 17, 12, 7, and 2. Formation 3 reaches the target with a single UAS loss occurring on edge $(13, 8)$. The path taken by formation 3 goes through nodes 24, 19, 18, 13, 8, 7, and 2. The UAS team lost three vehicles in total, and did not perform extra maneuvers when faced with the health threats.

Table 4.1 presents computing times and remaining maneuvering levels obtained with several values of $d(B_a, T)$ and d_m. The computer used for the simulations has a clock rate of 2 GHz and a RAM of 1 GB. For a given grid, when bound d_m is relaxed, the computing time rises significantly due to the increase in the number of feasible paths. Solving exactly DPE (4.16) is thus appropriate for a limited number of UAS formations, and for a limited number of maneuvering levels. Approximating DPE (4.16) with lookahead policies is one approach to constrain the computational burden; therefore, enabling a problem formulation with finer mesh granularity and larger number of maneuvering capability levels and formations. However, it must be emphasized that online routing and maneuvering management of the UAS with perfect knowledge of the environment, with application of the control signals obtained from the lookup table, is an extremely fast process. Figure 4.25 presents the typical paths obtained for a grid with $d_m = 15$, again for the case of three formations of three UAS. The paths are those actually followed by the ALTAVs. The models include the low-level dynamics and the GNC systems.

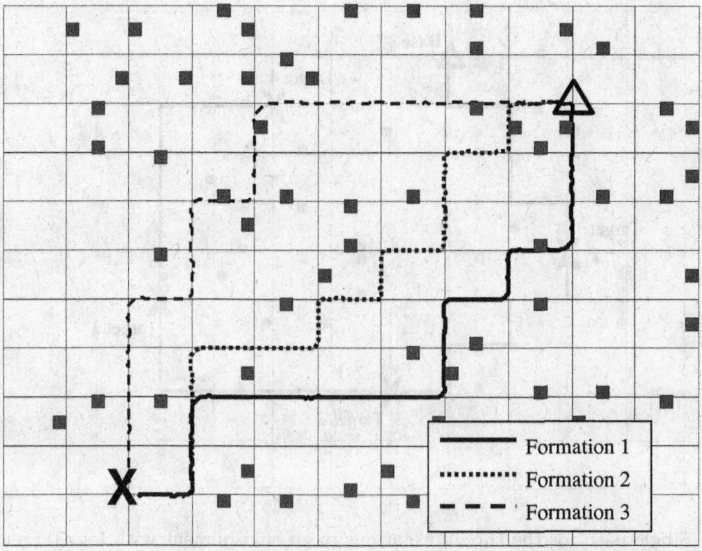

Fig. 4.25 Example of multi-formations for the case of $d_m = 15$.

4.3.2 *Sequence of targets and perfect knowledge of environment*

Consider the grid shown in Fig. 4.26. The UAS start from a base and have to reach targets 1, 2, 3 and 4. Figure 4.26 shows a single-path solution to the routing problem. For the multi-formations policy, Problem 4.1 is solved from B_a to T_1, and from T_1 to T_2, T_2 to T_3, and T_3 to T_4. In doing so, the constraint $d_m = 6$ is enforced between any two targets. The presence or absence of a threat along any given edge of the grid is determined probabilistically. Indeed, the probability that there is a threat along any given edge (i, j) is set to 0.55. The UAS-threat probabilities are the same as those of the previous section. All the threats are active and are of a high-level type.

It is assumed that the sequence of targets to be visited has been determined *a priori* by solving a multi-vehicle routing problem or one of its variants [155, 176]. This is the so-called formation-target assignment, or FTA, problem. RMM and FTA are integrated as shown in Fig. 4.27. Here, each formation is assigned the same sequence of targets. During UAS operation, the sequencing may be updated, if need be. For example, when

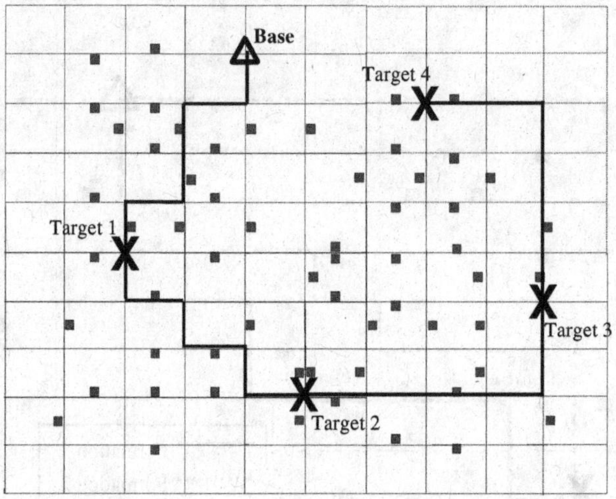

Fig. 4.26 Single path for the three formations in an environment with four targets.

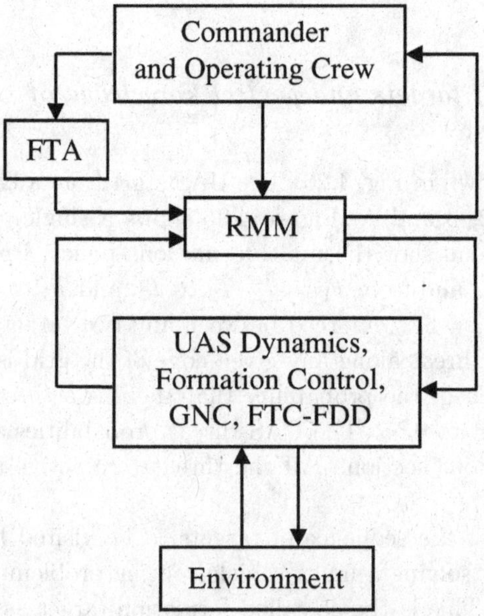

Fig. 4.27 Formation-target assignment, and routing and maneuvering management of cooperating UAS.

Table 4.2 Average number of maneuverability levels (AVG) and standard deviations (STD) for multi-formations path planning (RMM) and single path planning.

	T_1 (AVG)	T_2 (AVG)	T_3 (AVG, STD)	T_4 (AVG, STD)
RMM	17.8	12.8	9.7, 4.4	6.8, 4.3
Single path planning	13.8	7.7	3.8, 3.9	1.8, 1.6

Table 4.3 Empirical frequencies (%) on total number of simulations for which targets T_3 and T_4 are not reached at all.

	T_3 not reached	T_4 not reached
RMM	1.8	9.8
Single path planning	28.4	58.2

a significant event takes place, such as the occurrence of a pop-up threat and the displacement of a target, and when the commander issues new orders.

Figure 4.28 illustrates the optimal paths for the three formations. Performance indicators are given in Tables 4.2 and 4.3. As expected, the number of remaining maneuvering levels at each target is larger when the cooperating UAS are given the opportunity to divide and to aggregate (RMM) as opposed to being constrained to following a single path. Table 4.3 indicates the percentage of total simulation runs for which the multi-formations do not reach the last two targets. The third and fourth targets are very likely to be reached by at least one vehicle when equipped with the proposed RMM, that is more than 90% of the time, whereas the last target is reached by at least one vehicle with the single-routing policy for 41.8% of the total number of simulations. Figure 4.29 shows the remaining maneuvering levels at T_4 as a percentage of the number of simulation runs. Such graph can be used by a designer to determine the performance achievable for the cooperating UAS.

4.3.3 *The case of perturbed MDPs*

Consider again a group of nine UAS having to visit designated sectors T_1, T_2, T_3 and T_4, in that order. The simulations rely on the parameters described in the previous two sections. The threats are of high level. However, the simulations are run with perturbed MDPs whose (i, j)th entry

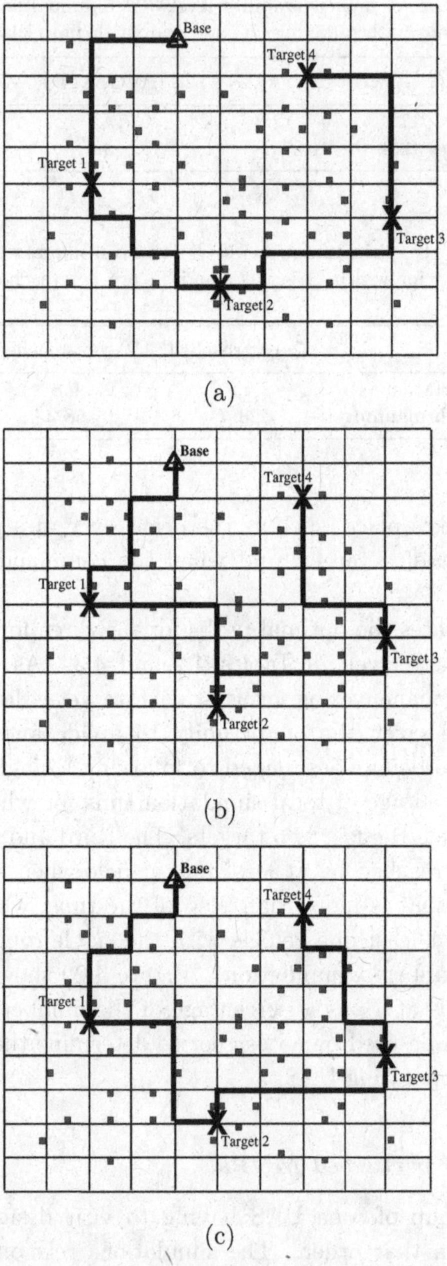

Fig. 4.28 Optimal paths for the planning of multi-formations: (a) formation 1, (b) formation 2, and (c) formation 3.

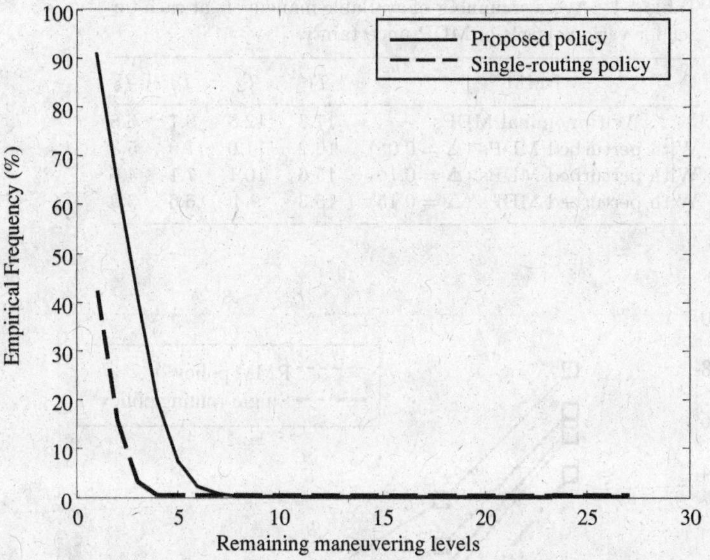

Fig. 4.29 Percentage of simulation runs versus remaining maneuvering levels at fourth target.

is defined as

$$\begin{aligned} p_{ij} &= p_{ij}^* + \delta_{ij} \\ &\geq 0 \end{aligned}$$

where p_{ij}^* is the (i, j)th entry of the nominal transition matrix (4.7), and p_{ij} is the perturbed probability. The following conditions must be satisfied

$$\sum_j p_{ij} = 1, \ \sum_j p_{ij}^* = 1$$

which imply that

$$\sum_j \delta_{ij} = 0.$$

The disturbance on the 2×2 MDP transition matrices is selected such that the actual probability of survival is decreased by $\Delta\%$ of its nominal value for the first row only; that is, $\delta_{11} = -\Delta \cdot p_{11}^*$. Then, $\delta_{21} = \Delta \cdot p_{21}^*$ to satisfy the above equalities. The DPE-based policy is calculated with nominal p_{ij}^*. However, the simulations are run with values of p_{ij} rather than the nominal p_{ij}^*. Parameter uncertainties are set to $\Delta = 0.05$, 0.1, and 0.15.

Table 4.4 Average number of available maneuvers at each target for various levels of MDP uncertainty.

RMM	T_1	T_2	T_3	T_4
With nominal MDPs	17.8	12.8	9.7	6.8
With perturbed MDPs ($\Delta = 0.05$)	16.2	11.0	7.9	5.3
With perturbed MDPs ($\Delta = 0.10$)	15.6	10.1	7.1	4.6
With perturbed MDPs ($\Delta = 0.15$)	15.3	9.4	6.6	3.9

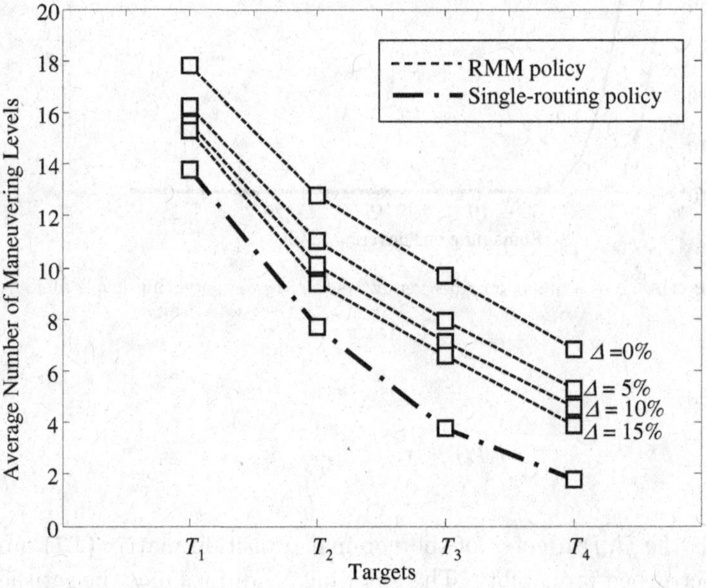

Fig. 4.30 Performance of policies as a function of uncertainty Δ.

The results of the simulations are given in Table 4.4. For $\Delta = 0.15$, there is a reduction in performance of up to 42.7%, when compared with simulations run with nominal MDPs, although the average number of available maneuvering levels obtained with RMM remains greater than that obtained with the single-routing policy, as shown in Fig. 4.30. The loss in performance due to the increase in uncertainty is not significant enough to incite the designer to replace the multi-formations strategy with the single-path planning. The results are context specific, however, and hence one cannot conclude that the performance of the RMM policy is always superior to that obtained with the single-routing policy.

4.3.4 *Successive zones of surveillance in partially known environment*

The cooperating UAS have to reach three target areas, or zones of surveillance, sequentially in a high-level threat environment, while a partial knowledge of the threat characteristics is available to the UAS team. The model of the environment is shown in Fig. 4.31. There are six uncertainties at the start of the mission. Each of those uncertainties corresponds either to a threat or to a safe edge (no threat). To calculate the base policy prior to mission, an active threat replaces each uncertainty. The presence of three SVs enables SIM-net to support the RMM of the cooperating UAS. Simulation parameters are the same as those presented in Sec. 4.3.1 with high-level threat probabilities given by Eqs. (4.44) and (4.45), although three targets have to be reached. To use the most recent information in the computation of a base policy for part of the theater, the lookup table required in planning the paths from target T_i to target T_{i+1} can be generated when the UAS are on their way to T_i, with uncertainties on the threats replaced by the worst-case situation: active threats.

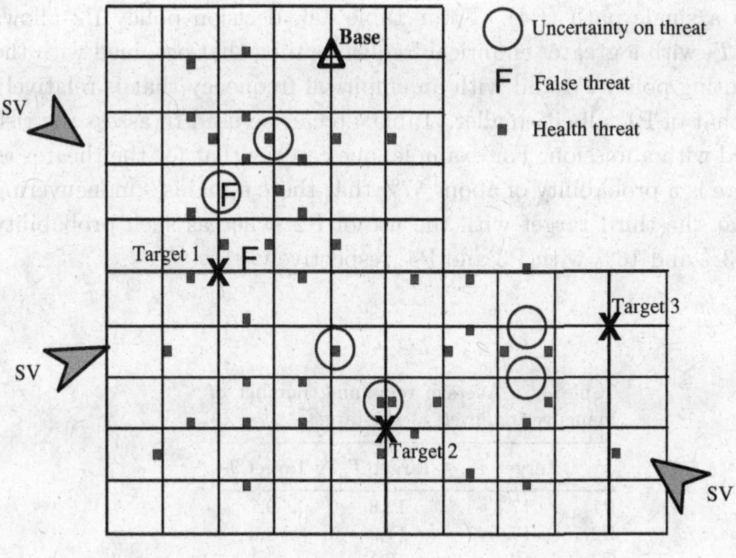

Fig. 4.31 Environment with three targets.

Five hundred simulation runs are performed. Due to the stochastic nature of the attrition dynamics, the path followed by the formations may change from one simulation to another. Thus, for brevity, results obtained by simulating the different policies are presented in Tables 4.5 and 4.6. The tables present the average remaining number of maneuvering levels at targets, and percentage of simulation runs yielding a certain number of remaining maneuvering levels.

Four policies are tested in the simulations. The base policy, denoted as P1, is computed and executed as if the environment is perfectly known and time invariant. This is the best possible scenario and serves as the yardstick against which are measured the performances of the other, more realistic, techniques and conditions of operation. Policy P2 is computed offline assuming an active threat is present in uncertain locations, and is updated online with the one-step lookahead policy improvement step. Policy P2 is simulated with imperfect knowledge of the environment, and is integrated with SIM-Net and the TC-information state RBF. Policy P3 is the base policy computed assuming an active threat is present in uncertain locations, although the policy is not updated online. Policy P4 is the single routing path planner. From Table 4.5, one can notice that the number of remaining maneuvering capability levels at each target is larger for policies P1, P2, and P3 as opposed to the case when formations are constrained to follow a single path (P4). From Table 4.6, decision policy P2 allows reaching T_3 with a greater empirical frequency than that obtained with the single-routing policy P4 and with an empirical frequency that is relatively close to that of P1, albeit smaller. Table 4.6 can be used to assess the risk associated with a mission. For example, one can say that for the theater at hand there is a probability of about 57% that there remains 9 maneuvering levels near the third target with the use of P2, whereas such probability falls to 43% and 16% with P3 and P4, respectively.

Table 4.5 Average remaining number of maneuvering levels at the targets.

	Target T_1	Target T_2	Target T_3
P1	17.8	12.8	9.7
P2	17.7	11.9	8.5
P3	16.1	10.1	7.1
P4	13.8	7.7	3.8

Table 4.6 Percentage of simulation runs yielding, at the third target, a certain number of remaining maneuvering levels (MLs) for the four decision policies (P1, P2, P3, P4).

	3 MLs	6 MLs	9 MLs	12 MLs	15 MLs	18 MLs	21 MLs
P1	95	87	65	41	21	8	3
P2	93	82	57	32	16	4	0
P3	90	72	43	20	8	2	0
P4	70	40	16	4	1	0	0

4.3.5 *Information flow faults*

Consider the mission described in the previous section and conducted within the high-level threat environment illustrated in Fig. 4.31. Furthermore, suppose that RMM-Net experiences intermittent communication breakdowns. The faults occur within time interval $[t_3, t_3 + \delta)$, namely when the formations fly from the second to the third node. Within interval $[t_3, t_3 + \delta)$, no inter-formation communication takes place. Recall that time t_1 corresponds to the start of the mission, with the UAS located at the base. An information flow fault prevents the formations from sharing their health state N_k, although the fault is not of sufficient duration to prevent the UAS from communicating their approximations on the expected cost-to-go function.

As before, five hundred simulation runs are carried out. Five policies are tested. P1 is the base policy computed and executed as if the environment is perfectly known, time-invariant, and exempt from faults. Policy P2 is the rollout policy derived and implemented under conditions of incomplete and partially known information on the TC and breakdown-free RMM-Net and SIM-Net communications. P2 is in closed loop with the TC-information state RBF. Policy P3 corresponds to policy P2 augmented with the N_k-information state RBF described in Sec. 4.2.3 to compensate for information flow faults of RMM-Net. Policy P4 corresponds to policy P2 in closed loop with an estimator of N_k, different from the N_k-information state RBF of P3. The estimator of P4 takes the last available health state information as the estimate. Finally, P5 is the single-formation, single-routing policy. Results are presented in Table 4.7. The realistic scenario characterized by the combination of imperfect information about the TC and intermittent communication losses results in a performance of P3 slightly inferior to that of P2. However, closing the loop with P3 results in a greater number of available maneuvering levels than that obtained with P4, which handles information loss in N_k by utilizing a data hold device, and with

Table 4.7 Average number of maneuvering capability levels remaining at each target.

	Target T_1	Target T_2	Target T_3
P1	17.8	12.8	9.7
P2	17.7	11.9	8.5
P3	17.5	11.6	8.5
P4	17.5	11.4	7.5
P5	13.8	7.7	3.8

the single-formation, single-routing policy P5. The results for P1 and P2, shown in Tables 4.5 and 4.7, are identical. Furthermore, policy P4 in Table 4.5 is the same as policy P5 in Table 4.7.

4.4 Distributed and Parallel Implementation of Optimization Algorithms

4.4.1 *Context*

The calculation of the optimal base policy is typically characterized by a high computational load, as discussed in Secs. 4.1 and 4.3. In particular, the figures given in Table 4.1 indicate that computing times become prohibitive as the dimensionality of the problem is increased. One solution to circumvent the computational complexity is to develop a parallel version of the dynamic programming equation. Several authors have investigated such approach. For instance, Refs. [193, 194] consider general models of asynchronous distributed computations, such as those found in communication networks, for which appropriate value-iteration algorithms are devised. Distributing the implementation of dynamic programming algorithms has also been investigated in the contexts of reconfigurable networks [195], clusters of workstations with serial monadic algorithms [196], and computational biology [197], to cite a few. We propose in the sequel a synchronous distributed version of the DPE, which is aimed at yielding fast computation of the base policy and cost-to-go function in (4.16) expressed as lookup tables. This distributed implementation of the DPE is tested on a COTS network of personal computers, such as the RT-LAB® software/hardware platform [198], where the modeling of the various systems is done in Simulink® [199].

4.4.2 *Architecture*

The solution to DPE (4.16) depends on state N_j at iteration j and on configuration $c_{c,j}$. The total number of states of the formations and possible configurations are $(n_p+1)^p$ and m_j^p, respectively, where m_j represents the number of nodes of E_j. Recall that p is the number of formations at the start of the mission. The reader is referred to Sec. 4.1 for more details on those variables. A convenient way to accelerate the computation of the policy obtained from DPE (4.16) is to distribute the recursion of the DPE over a network of $l \in \mathbb{N}$ computers. Each computer implements a parallel formulation of DPE (4.16) labeled as a subDPE. A subDPE is calculated for a subset of the m_j^p possible configurations $c_{c,j}$. The $(n_p+1)^p$ possible values of states N_j are integrally kept in the computations, as required in Eq. (4.16), which involves every possible value of N_j. The subDPE implemented on computer $\mathcal{C} \in \{1,...,l\}$, can be expressed, at iteration $j \in \{1,...,N-1\}$ and for all

$$c' \in \{(\mathcal{C}-1)\frac{m_j^p}{l}+1,...,\mathcal{C}\frac{m_j^p}{l}\} = S_{\mathcal{C}},$$

as

$$\begin{aligned} V_{j,P_i,\gamma^+}(N_j, c_{c',j}) = \min_{\substack{U_j \in \mathcal{U}_j(r_{c_{d,j}}) \\ r_{c_{d,j}} \in R_{c_{c',j}}}} \max_{V_j} (\textstyle\sum_{\nu \in B} U_{2,j}^{\nu} \\ + E_{\substack{\gamma^+ \\ N_{j+1}}} \{\textstyle\sum_{\nu \in B} m_p I(N_{j+1}^{\nu} | N_j^{\nu}) + V_{j+1,P_i,\gamma^+}(N_{j+1}, c_{d,j+1})\}), \end{aligned} \tag{4.46}$$

where $d \in \{1,...,m_j^p\}$. We assume that the elements of $S_{\mathcal{C}}$ are integers. The scheduling of the computing tasks distributed over the network of computers is shown in Fig. 4.32 and is explained as follows. At iteration j, every subDPE first reads the cost-to-go functions

$$V_{j+1,P_i,\gamma^+}(N_{j+1}, c_{d',j+1}),$$

with $d' \in S_{\mathcal{C}}$, as communicated by the other computers. Second, the $(n_p+1)^p m_j^p/l$ cost-to-go functions, instead of the $(n_p+1)^p m_j^p$ functions of a centralized implementation, are updated by means of Eq. (4.46). Intimately linked to the computation of the cost-to-go function is the solution to the min-max problem which yields U_j and V_j. Every subDPE then outputs $V_{j,P_i,\gamma^+}(N_j, c_{c',j})$ for all $c' \in S_{\mathcal{C}}$ which are read by the other computers at the next iteration (or time) step.

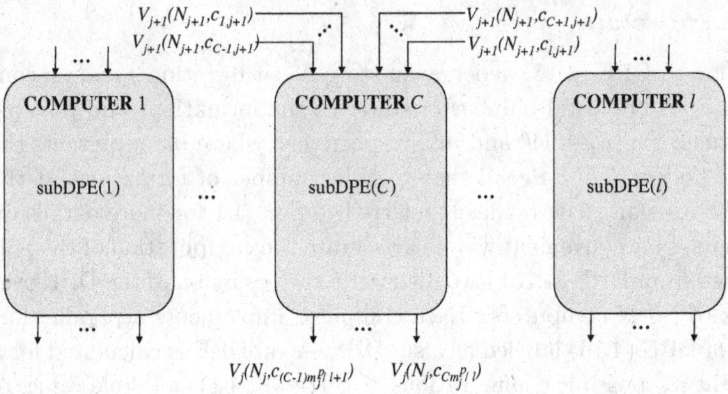

Fig. 4.32 Distributed implementation of the computation of the base policy obtained with DPE (4.16).

4.4.3 *Distributed and parallel simulation environment*

Programming subDPE (4.46) can be readily done in a numerical simulation environment such as MATLAB®. This software is adopted by the authors for its optimized matrix calculus, its efficient debugging interface, its ease of use for implementing and simulating dynamic systems, and for its automatic code generation capability. Furthermore, the existence of software and hardware platforms to distribute the code obtained from a Simulink® model unto a distributed or a parallel architecture enables us to adopt a top-down approach for programming, debugging and generating real-time code within a common environment [198, 200]. Such approach allows reducing the design and implementation times while minimizing the risk of corrupting the original concepts in the process. We label such capability as rapid prototyping for cooperative control.

The conceptual distributed computation of the base policy depicted in Fig. 4.32 can be actually implemented with the architecture shown in Fig. 4.33, for a four-computer network. Each computing node executes the code obtained from a Simulink® model of subDPE (4.46). Simulink® models are typically represented by means of discrete-time dynamic systems given, at iteration k, as

$$\begin{aligned} x_{k+1} &= f(x_k, u_k), \\ y_k &= h(x_k, u_k), \end{aligned} \tag{4.47}$$

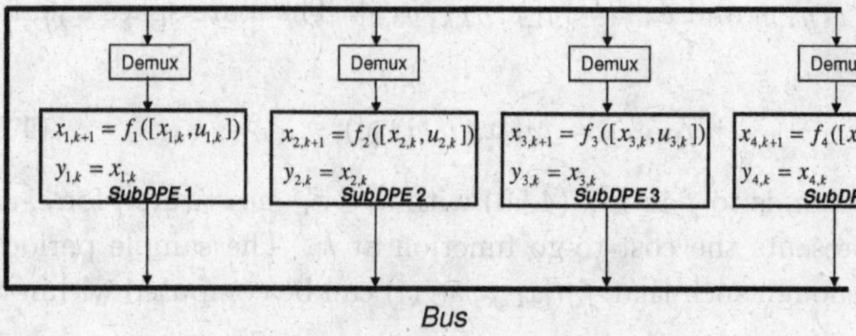

Fig. 4.33 Implementation of distributed DPE on a four-computer network.

where x_k, u_k, and y_k stand for the state, input, and output variables, respectively. Letting

$$\begin{aligned} x_0 &= V_{N_N, P_i, \gamma^+}(N_N, T), \\ x_k &= V_{N-k, P_i, \gamma^+}(N_{N-k}, c_{d,N-k}), \end{aligned} \tag{4.48}$$

yields the discrete-time system in Eqs. (4.47), with

$$\begin{aligned} u_k &= x_k, \\ y_k &= x_k, \end{aligned} \tag{4.49}$$

and

$$\begin{aligned} f(u_k) = \min_{\substack{U_{k+1} \in \mathcal{U}_{k+1}(r_{c_{d,k+1}}) \\ r_{c_{d,k+1}} \in R_{c_{c,k+1}}}} & \max_{V_k}(\textstyle\sum_{\nu \in B} A \cdot U^{\nu}_{2,k+1} \\ & + E_{\substack{\gamma^+ \\ N_k}}\{\textstyle\sum_{\nu \in B} m_p I(N^{\nu}_k | N^{\nu}_{k+1}) + u_k\}). \end{aligned} \tag{4.50}$$

In Eq. (4.50), explicit dependency on x_k is dropped since $u_k = x_k$. $U^{\nu}_{2,k+1}$ and N^{ν}_k are considered as endogenous to f, and hence are not state variables of f. The state-space representation of the subDPE does not exhibit a direct feedthrough; that is, y_k does not depend on u_k. The processes are therefore exempt from algebraic loops. Distributing the code is a straightforward process, as shown in Fig. 4.33 for a network of four computers. A subDPE is uniquely associated with a computer in the network. Each computer $\mathcal{C}$ communicates information to the other computers by means of a bus which includes variables $y_{1,k}$, $y_{2,k}$, $y_{3,k}$, and $y_{4,k}$. A demultiplexer, labeled as Demux in the figure, selects signals $y_{i,k}$ where $i \in \{1,2,3,4\} \backslash \mathcal{C}$. In other words, the input to computer 1 is $u_{1,k} = [y_{2,k}, y_{3,k}, y_{4,k}]$, the input to computer 2 is $u_{2,k} = [y_{1,k}, y_{3,k}, y_{4,k}]$, whereas the other inputs are

$u_{3,k} = [y_{1,k}, y_{2,k}, y_{4,k}]$ and $u_{4,k} = [y_{1,k}, y_{2,k}, y_{3,k}]$. The state-space equation of C is given by

$$x_{C,k+1} = f_C([x_{C,k}, u_{C,k}]), \tag{4.51}$$

where f_C corresponds to f in Eq. (4.50) with $c' \in S_C$ substituted for c, and $[x_{C,k}, u_{C,k}]$ represents the cost-to-go function at k. The sample period is selected large enough such that $f_C([x_{C,k}, u_{C,k}])$ can be computed within the sample period.

4.4.4 *Experiments*

The calculation of the base policy is distributed on two computers, with an architecture shown in Fig. 4.34. Offline design and analysis are performed on the host personal computer (PC). User commands are transmitted from the host PC to the RT targets through a TCP/IP link. The real-time processing environment comprises two central processing units, or RT targets, running RedHawk Linux real-time operating system [201]. Each target is connected via TCP/IP to the host and viewer PC. The RT targets share information on the cost-to-go function through a communication network, for example serial communication IEEE 1394 (FireWire), ethernet, or simply shared memory among multiple processors, depending on available hardware. Designer-specified data are communicated to the viewer PC equipped with a renderer software to display the vehicles and the environment, if need be. With the availability of a nominal Simulink® model, the designer separates the various components of his/her model into subsystems, such as Eqs. (4.46)-(4.51). With RT-LAB® [198], the designer assigns a subDPE, modeled as a subsystem in Simulink®, to any of the RT targets. Then, the designer initiates the process of automatic code generation, compilation on the RT targets, and uploading the computing processes associated with the subDPE to the assigned RT targets. Once such process is done, the user can run the simulations and acquire the result of the computation of the base policy. Theoretically, we expect that the original DPE divided into N subDPE results in computing times which are also reduced by a factor of N.

In this book, experiments are confined to the case $N = 2$. The targets are two 2-GHz computers having 1 GB RAM. The targets communicate information by means of shared memory. For a scenario with $d_m = 6$, a single-target computation takes 10 seconds, whereas a 2-target implementation requires 5.2 and 6 seconds on the respective processing units. In

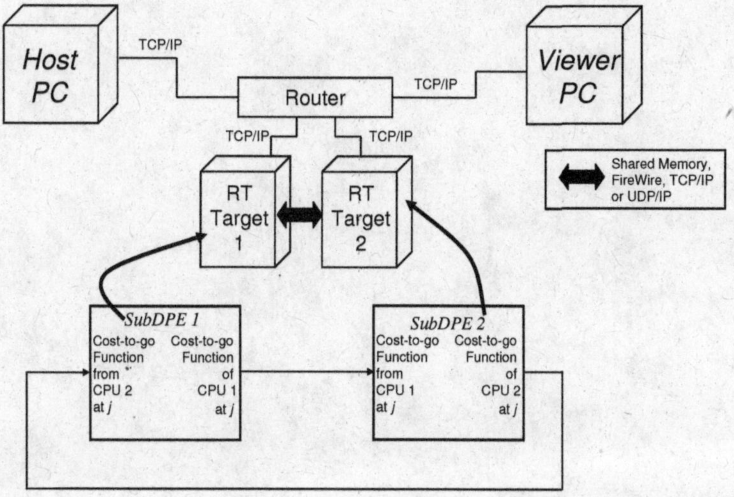

Fig. 4.34 Technological platform for the distribution on two computers.

such case, the reduction factor in computing time is 1.67. A scenario with $d_m = 8$ necessitates 82 seconds of computation for a single processor implementation, as opposed to 61 and 62 seconds for a 2-processor distribution. A reduction in computing time by a factor of 1.32 is thus obtained. The experiments indicate that a distributed computation of the DPE on a local network of COTS personal computers reduces the time needed to obtain the base policy. The reduction in computing time is however not as significant as theoretically expected for the limited set of cases implemented. Communication overheads and software/hardware synchronization issues have an influence on the performance. Naturally, the concept of distributed DPE can be extended to other, larger networks of computing devices, such as the network of UAS, provided the processing and communication capabilities allow for such distributed computations. Distributed DPE computations differ from conventional time-based simulations on two fronts. On the one hand, the number of iterations (or time steps) needed to solve the DPE is relatively small, depending on the value of d_m. On the other hand, the size of the matrices transmitted from one subDPE to another may become prohibitively large, for the technological medium available, depending on the dimensionality of the problem. The dimensionality increases with the difference between the actual path length and d_m.

Bibliography

[1] Van Cleave, D. W. (2003). *Trends and Technologies for Unmanned Aerial Vehicles*, ch. in *Software-Enabled Control*, Samad, T. and Balas, G. (eds.) (IEEE Press).

[2] United States Office of the Secretary of Defense (2007). *Unmanned Systems Roadmap 2007-2032.*

[3] Zufferey, J.-C. and Floreano, D. (2006). Fly-Inspired Visual Steering of an Ultralight Indoor Aircraft, *IEEE Transactions on Robotics*, Vol. 22, No. 1, pp. 137–146.

[4] Shima, T., Rasmussen, S., and Gross, D. (2007). Assigning Micro UAVs to Task Tours in an Urban Terrain, *IEEE Transactions on Control Systems Technology*, Vol. 15, No. 4, pp. 601–612.

[5] Kumar, V., Leonard, N. and Morse, A. S. (eds.) (2005). *Cooperative Control* (Springer-Verlag).

[6] Pettersen, K. Y., Gravdahl, J. T., and Nijmeijer, H. (eds.) (2006). *Group Coordination and Cooperative Control* (Springer-Verlag).

[7] Grundel, D., Murphey, R., Pardalos, P., and Prokopyev, O. (eds.) (2007) *Cooperative Systems* (Springer-Verlag).

[8] Butenko, S., Murphey, R., and Pardalos, P. (eds.) (2004) *Recent Developments in Cooperative Control and Optimization* (Kluwer Academic Publishers).

[9] United States Office of the Secretary of Defense (2005). *Unmanned Aircraft Systems Roadmap 2005-2030.*

[10] Castillo, P., Lozano, R., and Dzul, A. E. (2005) *Modelling and Control of Mini-Flying Machines* (Springer).

[11] Wu, H., Sun, D., and Zhou, Z. (2004). Micro Air Vehicle: Configuration, Analysis, Fabrication, and Test, *IEEE/ASME Transactions on Mechatronics*, Vol. 9, No. 1, pp. 108–117.

[12] Green, W. E. and Oh, P. Y. (2005). A MAV That Flies Like an Airplane and Hovers Like a Helicopter, in *Proceedings of the 2005 IEEE/ASME International Conference on Advanced Intelligent Mechatronics*, Monterey, California, pp. 693–698.

[13] Mettler, B., Dever, C., and Feron, E. (2004). Scaling Effects and Dynamic

Characteristics of Miniature Rotorcraft, *Journal of Guidance, Control and Dynamics*, Vol. 27, No. 3, pp. 466–478.

[14] Léchevin, N., Rabbath, C. A., and Earon, E. (2007). Towards Decentralized Fault Detection in UAV Formations, in *Proceedings of American Control Conference 2007*, New York, NY, pp. 5759–5764.

[15] Léchevin, N., and Rabbath, C. A. (2007). Robust Decentralized Fault Detection in Leader-to-Follower Formations of Uncertain, Linearly Parameterized Systems, *Journal of Guidance, Control, and Dynamics*, Vol. 30, No. 5, pp. 1528–1535.

[16] Bouabdallah, S., Becker, M., and Siegwart, R. (2007). Autonomous Miniature Flying Robots: Coming Soon, *IEEE Robotics and Automation Magazine*, September, pp. 88–98.

[17] Murphey, R. A. (2002). *Cooperative Control and Optimization*, ch. in *An Introduction to Collective and Cooperative Systems* (Kluwer).

[18] Giulietti, F. , Pollini, L., and Innocenti, M. (2000). Autonomous Formation Flight, *IEEE Control Systems Magazine*, December, pp. 34–44.

[19] Boskovic, J. D., Bergstrom, S., and Mehra, R. K. (2005). Robust Integrated Flight Control Design Under Failures, Damage, and State-Dependent Disturbances, *Journal of Guidance, Control, and Dynamics*, Vol. 28, No. 5, pp. 902–917.

[20] Tao, G., Chen, S., Tang, X., and Joshi, S. M. (2004). *Adaptive Control of Systems with Actuator Failures* (Springer-Verlag).

[21] Chen, J., and Patton, R. J. (1999). *Robust Model-Based Fault Diagnosis for Dynamic Systems* (Kluwer Academic Publishers).

[22] Napolitano, M. R., Windon, D. A., II, Casanova, J. L., Innocenti, M., and Silvestri, G. (1998). Kalman Filters and Neural-Network Schemes for Sensor Validation in Flight Control Systems, *IEEE Transactions on Control Systems Technology*, Vol. 6, No. 5, September, pp. 596–611.

[23] Boskovic, J. D. and Mehra, R. K. (2002). Stable Adaptive Multiple Model-based Control Design for Accommodation of Sensor Failures, in *Proceedings of the American Control Conference*, Anchorage, AK, pp. 2046–2051.

[24] Samad, T., Bay, J. S., and Godbole, D. (2007). Network-Centric Systems for Military Operations in Urban Terrain: The Role of UAVs, *Proceedings of the IEEE*, Vol. 95, No. 1, pp. 92–107.

[25] Ren, W., Chao, H., Bourgeous, W., Sorensen, N., and Chen, Y. Q. (2008). Experimental Validation of Consensus Algorithms for Multivehicle Cooperative Control, *IEEE Transactions on Control Systems Technology*, Vol. 16, No. 4, pp. 745–752.

[26] Boskovic, J. D., Bergstrom, S. E., and Mehra, R. K. (2005). Retrofit Reconfigurable Flight Control in the Presence of Control Effector Damage, in *Proceedings of 2005 American Control Conference*, Portland, OR, pp. 2652–2657.

[27] Léchevin, N., and Rabbath, C. A. (2009). Decentralized Detection of a Class of Nonabrupt Faults With Application to Formations of Unmanned Airships, *IEEE Transactions on Control Systems Technology*, Vol. 17, No. 2, pp. 484–493.

[28] Kumar, P. R. (2001). New Technological Vistas for Systems and Control: The Example of Wireless Networks, *IEEE Control Systems Magazine*, February, pp. 24–37.

[29] Jagannathan, S. (2007). *Wireless Ad Hoc and Sensor Networks - Protocols, Performance, and Control* (CRC Press).

[30] Khatib, O. (1986). Real-time obstacle avoidance for manipulators and mobile robots, *International Journal of Robotics Research*, Vol. 5, No. 1, pp. 90–98.

[31] Tanner, H. G., and Christodoulakis, D. K. (2007). Decentralized cooperative control of heterogeneous vehicle groups, *Robotics and Autonomous Systems*, Vol. 55, No. 11, pp. 811–823.

[32] Shim, D. H., Chung, H., and Sastry, S. S. (2006). Conflict-Free Navigation in Unknown Urban Environments, *IEEE Robotics and Automation Magazine*, September, pp. 27–33.

[33] Khalil, H. K. (2002). *Nonlinear Systems*, (Prentice Hall).

[34] Lohmiller, W., and Slotine, J.-J. E. (1998). On contraction analysis for nonlinear systems, *Automatica*, Vol. 34, No. 6, pp. 683–696.

[35] Feinberg, E. A., and Schwartz, A. (2002). *Handbook of Markov Decision Process - Methods and Applications* (Kluwer).

[36] Bertsekas, D. (2005). *Dynamic Programming and Optimal Control- Volume I* (Athena Scientific).

[37] Zhang, Y. M., and Jiang, J. (2003). Bibliographical review on reconfigurable fault-tolerant control systems, in *Preprints of the 5th IFAC Symp. on Fault Detection, Supervision and Safety of Technical Processes*, Washington, DC, USA, pp. 265–276.

[38] Forssell, L., and Nilsson, U. (2005). ADMIRE the Aero-Data Model in a Research Environment Version 4.0, Model Description, Technical report FOI-R-1624-SE.

[39] Bates, D. and Postlethwaite, I. (2002). *Robust Multivariable Control of Aerospace Systems* (Delft University Press).

[40] Eterno, J. S., Weiss, J. L., Looze, D. P., and Willsky, A. S. (1985). Design issues for fault tolerant-restructurable aircraft control, in *Proc. of the 24th IEEE Conference on Decision and Control*, Fort Lauderdale, FL, pp. 900–905.

[41] Zhang, Y. (2006). Current Status and Future Trends in Fault Tolerant Control, *Workshop on Advanced Control and Diagnosis*, Nancy, France.

[42] Liao, F., Wang, J. L., and Yang, G.-H. (2002). Reliable Robust Flight Tracking Control: an LMI Approach, *IEEE Transactions on Control Systems Technology*, Vol. 10, No. 1, pp. 76–89.

[43] Gertler, J. J. (1998). *Fault Detection and Diagnosis in Engineering Systems* (Marcel Dekker).

[44] Jiang, J. and Zhang, Y. M. (2006). Accepting Performance Degradation in Fault-Tolerant Control System Design, *IEEE Transactions on Control Systems Technology*, Vol. 14, No. 2, pp. 284–292.

[45] Frank, P. M. (1990). Fault diagnosis in dynamic systems using analytical

and knowledge-based redundancy - a survey and some new results, *Automatica*, Vol. 26, No. 3, pp. 459–474.

[46] Blanke, M., Kinnaert, M., Lunze, J., and Staroswiecki, M. (2003). *Diagnosis and Fault-Tolerant Control* (Springer).

[47] Massoumnia, M.-A. (1986). A Geometric Approach to the Synthesis of Failure Detection Filters, *IEEE Transactions on Automatic Control*, Vol. 31, pp. 839–846, 1986.

[48] Zhang, Y. M., and Jiang, J. (2001). Integrated active fault-tolerant control using IMM approach, *IEEE Transactions on Aerospace and Electronic Systems*, Vol. 37, No. 4, pp. 1221–1235.

[49] Ducard, G., and Geering, H. P. (2008). Efficient Nonlinear Actuator Fault Detection and Isolation System for Unmanned Aerial Vehicles, *Journal of Guidance, Control, and Dynamics*, Vol. 31, No. 1, pp. 225–237.

[50] Low, X. C., Willsky, A. S., and Verghese, G. L. (1986). Optimally robust redundancy relations for failure detection in uncertain systems, *Automatica*, Vol. 22, pp. 333–344.

[51] Chow, E. Y., and Willsky, A. S. (1984). Analytical Redundancy and the Design of Robust Failure Detection Systems, *IEEE Transactions on Automatic Control*, Vol. 29, No. 7, pp. 603–614.

[52] Zhang, Y. M., and Jiang, J. (2001). Integrated Design of Reconfigurable Fault-tolerant Control Systems, *Journal of Guidance, Control, and Dynamics*, Vol. 24, No. 1, pp. 133–136.

[53] Shen, L.-C., Chang, S.-K., and Hsu, P.-L. (1998). Robust Fault Detection and Isolation with Unstructured Uncertainty Using Eigenstructure Assignment, *Journal of Guidance, Control, and Dynamics*, Vol. 21 No. 1, pp. 50–57.

[54] Chen, B. and Nagarajaiah, S. (2007). Linear-Matrix-Inequality-Based Robust Fault Detection and Isolation Using the Eigenstructure Assignment Method, *Journal of Guidance, Control, and Dynamics*, Vol. 30, No. 6, pp. 1831–1835.

[55] Jiang, B., Staroswiecki, M., and Cocquempot, V. (2004). Fault estimation in nonlinear uncertain systems using robust/sliding-mode observers, *IEE Proceedings - Control Theory and Applications*, Vol. 151, No. 1, pp. 29–37.

[56] Seliger, R., and Frank, P. M. (1991). Fault diagnosis by disturbance decoupled nonlinear observer, in *Proceedings of the IEEE Conference on Decision and Control*, 1991, pp. 2248–2253.

[57] Persis, C. D., and Isidori, A. (2001). A Geometric Approach to Nonlinear Fault Detection and Isolation, *IEEE Transactions on Automatic Control*, Vol. 46, No. 6, pp. 853–865.

[58] Polycarpou, M. M., and Helmicki, M.-A. (1995). Automated fault detection and accommodation: A learning approach, *IEEE Transactions on Systems, Man, and Cybernetics*, Vol. 25, No. 11, pp. 1447–1458.

[59] Zhang, X., Parisini, T., and Polycarpou, M. M. (2004). Adaptive fault-tolerant control of nonlinear uncertain systems: an information based diagnostic approach, *IEEE Transactions on Automatic Control*, Vol. 49, No. 8, pp. 1259–1274.

[60] Perhinschi, M. G., Napolitano, M. R., Campa, G., Seanor, B., Burken, J., and Larson, R. (2006). An Adaptive Threshold Approach for the Design of an Actuator Failure Detection and Identification Scheme, *IEEE Transactions on Control Systems Technology*, Vol. 14, No. 3, pp. 519–525.

[61] Zhang, Y. M., and Jiang, J. (2002). Design of Restructurable Active Fault-tolerant Control Systems, in *Preprints of the 15th IFAC World Congress*, Barcelona, Spain.

[62] Looze, D., Weiss, J., Eterno, J., and Barrett, N. (1985). An automatic redesign approach for restructurable control systems, *IEEE Control Systems Magazine*, Vol. 5, No. 2, pp. 16–22.

[63] Gao, Z. and Antsaklis, P. J. (1991). Stability of the Pseudo-inverse Method for Reconfigurable Control Systems, *International Journal of Control*, Vol. 53, No. 3, pp. 717–729.

[64] Bacon, B. J., Ostroff, A. J., and Joshi, S. M. (2001). Reconfigurable NDI Controller Using Inertial Sensor Failure Detection and Isolation, *IEEE Transactions on Aerospace and Electronic Systems*, Vol. 37, No. 4, pp. 1373–1383.

[65] Gopinathan, M., Boskovic, J., Mehra, R. K., and Rago, C. (1998). A Multiple Model Predictive Scheme for Fault-tolerant Flight Control Design, in *Proceedings of the 37th IEEE Conference on Decision and Control*, Tampa, FL, pp. 1376–1381.

[66] Luo, Y., Serrani, A., Yurkovich, S., Oppenheimer, M. W., and Doman, D. B. (2007). Model-Predictive Dynamic Control Allocation Scheme for Reentry Vehicles, *Journal of Guidance, Control, and Dynamics*, Vol. 30, No. 1, pp. 100–113.

[67] Napolitano, M. R., An, Y., and Seanor, B. A. (2000). A Fault Tolerant Flight Control System for Sensor and Actuator Failures Using Neural Networks, *Aircraft Design*, Vol. 3, No. 2, pp. 103–128.

[68] Shin, D.-H. and Kim, Y. (2004). Reconfigurable flight control system design using adaptive neural networks, *IEEE Transactions on Control Systems Technology*, Vol. 12, No. 1, pp. 87–100.

[69] Suresh, S., Omkar, S. N., and Mani, V. (2006). Direct Adaptive Neural Flight Controller for F-8 Fighter Aircraft, *Journal of Guidance, Control, and Dynamics*, Vol. 29, No. 2, pp. 454–464.

[70] Cieslak, J., Henry, D., Zolghadri, A., and Goupil, P. (2008). Development of an Active Fault-Tolerant Flight Control Strategy, *Journal of Guidance, Control, and Dynamics*, Vol. 31, No. 1, pp. 135–147.

[71] Zhang, Y. and Jiang, J. (2001). Integrated active fault-tolerant control using IMM approach, *IEEE Transactions on Aerospace and Electronic Systems*, Vol. 37, No. 4, pp. 1221–1235.

[72] Jiang, J. (1994). Design of Reconfigurable Control Systems Using Eigenstructure Assignment, *International Journal of Control*, Vol. 59, No. 2, pp. 395–410.

[73] Huang, C.Y. and Stengel, R.F. (1990). Restructurable Control Using Proportional-Integral Implicit Model Following, *Journal of Guidance, Control, and Dynamics*, Vol. 13, No. 2, pp. 303–309.

[74] Maybeck, P. S. and Stevens, R. D. (1991). Reconfigurable flight control via multiple model adaptive control methods, *IEEE Transactions on Aerospace and Electronic Systems*, Vol. 27, No. 3, pp. 470–480.
[75] Bodson, M. and Groszkiewicz, J. E. (1997). Multivariable adaptive algorithms for reconfigurable flight control, *IEEE Transactions on Control Systems Technology*, Vol. 5, No. 2, pp. 217–229.
[76] Ward, D. G., Monaco, J. F., and Bodson, M. (1998). Development and Flight Testing of a Parameter Identification Algorithm for Reconfigurable Control, *Journal of Guidance, Control, and Dynamics*, Vol. 21, No. 6, pp. 948–956.
[77] Burken, J.J., Lu, P., Wu, Z., and Bahm, C. (2001). Two Reconfigurable Flight-Control Design Methods: Robust Servomechanism and Control Allocation, *Journal of Guidance, Control, and Dynamics*, Vol. 24, No. 3, pp. 482–493.
[78] Härkegård, O. (2004). Dynamic Control Allocation Using Constrained Quadratic Programming, *Journal of Guidance, Control, and Dynamics*, Vol. 27, No. 6, pp. 1028–1034.
[79] Liu, Y., Tang, X., Tao, G., and Joshi, S. M. (2008). Adaptive Compensation of Aircraft Actuation Failures Using an Engine Differential Model, *IEEE Transactions on Control Systems Technology*, Vol. 16, No. 5, pp. 971–982.
[80] Guler, M., Clements, S., Wills, L. M., Heck, B. S., and Vachtsevanos, G. J. (2003). Transition Management for Reconfigurable Hybrid Control Systems, *IEEE Control Systems Magazine*, February, pp. 36–49.
[81] Blanke, M., Staroswiecki, M., and Wu, N. E. (2001). Concepts and methods in fault-tolerant control, in *Proc. of the 2001 American Control Conference*, Arlington, USA, pp. 2606–2620.
[82] Guler, M., Clements, S., Kejriwal, N., Wills, L., Heck, B., Vachtsevanos, G. (2002). Rapid Prototyping of Transition Management Code for Reconfigurable Control Systems, in *Proc. of the 13th IEEE Int. Workshop on Rapid Systems Prototyping*, Darmstadt, Germany, pp. 76–83.
[83] Wills, L., Kannan, S., Sander, S., Guler, M., Heck, B., Prasad, J. V. R., Schrage, D., and Vachtsevanos, G. (2001). An Open Platform for Reconfigurable Control, *IEEE Control Systems Magazine*, June, pp. 49–64.
[84] Orr, M. W., Rasmussen, S. J., Karni, E. D., and Blake, W. B. (2005). Framework for Developing and Evaluating MAV Control Algorithms in a Realistic Urban Setting, in *Proceedings of 2005 American Control Conference*, Portland, OR, pp. 4096–4101.
[85] Bethke, B., Valenti, M., and How, J. P. (2008). UAV Task Assignment - An Experimental Demonstration with Integrated Health Monitoring, *IEEE Robotics and Automation Magazine*, March, pp. 39–44.
[86] Mohr, B. B. and Fitzpatrick, D. L. (2008). Micro Air Vehicle Navigation System, *IEEE Aerospace and Electronic Systems Magazine*, pp. 19–24.
[87] How, J. P., Bethke, B., Frank, A., Dale, D., and Vian, J. (2008). Real-Time Indoor Autonomous Vehicle Test Environment, *IEEE Control Systems Magazine*, April, pp. 51–64.
[88] Boskovic, J. D., Prasanth, R., and Mehra, R. K. (2004). A Multi-Layer

Autonomous Intelligent Control Architecture for Unmanned Aerial Vehicles, *Journal of Aerospace Computing, Information, and Communication*, Vol. 1, December, pp. 605–628.

[89] Slegers, N., and Costello, M. (2007). Variable Structure Observer for Control Bias on Unmanned Air Vehicles, *Journal of Guidance, Control, and Dynamics*, Vol. 30, No. 1, pp. 281–286.

[90] Shore, D., and Bodson, M. (2005). Flight Testing of a Reconfigurable Control System on an Unmanned Aircraft, *Journal of Guidance, Control, and Dynamics*, Vol. 28, No. 4, pp. 698–707.

[91] Johnson, E. N., and Schrage, D. P. (2004). System Integration and Operation of a Research Unmanned Aerial Vehicle, *Journal of Aerospace Computing, Information, and Communication*, Vol. 1, No. 1, pp. 5–18.

[92] Basseville, M. and Nikiforov, I. V. (1993). *Detection of Abrupt Changes* (Prentice-Hall).

[93] Boskovic, J. D., Bergstrom, S. E., and Mehra, R. K. (2005). Robust Integrated Control Design Under Failures, Damage, and State-Dependant Disturbances, *AIAA Journal of Guidance, Control, and Dynamics*, Vol. 28, No. 5, pp. 902–917.

[94] Rogge, J.A. and Aeyels, D. (2008). Vehicle Platoons Through Ring Coupling, *IEEE Tansactions on Automatic Control*, Vol. 53, No. 6, pp. 1370–1377.

[95] Dong, W., and Farrell, J. A. (2008). Cooperative Control of Multiple Nonholonomic Mobile Agents, *IEEE Tansactions on Automatic Control*, Vol. 53, No. 6, pp. 1434–1438.

[96] Do, K.D. (2008). Formation Tracking Control of Unicycle-Type Mobile Robots With Limited Sensing Ranges, *IEEE Tansactions on Control Systems Technology*, Vol. 16, No. 3, pp. 527–538.

[97] Jadbabaie, A., Lin, J., and Morse, A. S. (2003). Coordination of groups of mobile autonomous agents using nearest neighbors rules, *IEEE Transactions on Automatic Control*, Vol. 48, No. 6, pp. 988–1001.

[98] Slotine, J.-J. E. and Wang, W. (2005) *A study of synchronization and group cooperation using partial contraction theory*, ch. in *Cooperative Control LNCIS 309* (Springer).

[99] Marshall, J. A., Broucke, M. E., and Francis, B. A. (2004). Formations of vehicles in cyclic pursuit, *IEEE Transactions on Automatic Control*, Vol. 49, No. 11, pp. 1963–1974.

[100] Moreau, L. (2005). Stability of multiagent systems with time-dependent communication links, *IEEE Transactions on Automatic Control*, Vol. 50, No. 2, pp. 169–182.

[101] Lawton, J. R. T., Beard, R. W., and Young, Y. (2003). A decentralized approach to formation maneuvers, *IEEE Transactions on Robotics and Automation*, Vol. 19, No. 6, pp. 933–941.

[102] Pant, A., Seiler, P., and Hedrick, K. (2002). Mesh stability of look-ahead interconnected systems, *IEEE Transactions on Automatic Control*, Vol. 47, No. 2, pp. 403–407.

[103] Pant, A., Seiler, P., and Hedrick, K. (2004). Disturbance Propagation in

Vehicle Strings, *IEEE Transactions on Automatic Control*, Vol. 49, No. 10, pp. 1835–1841.

[104] Liu, Y., Passino, K. M., and Polycarpou, M. (2003). Stability analysis of one-dimensional asynchronous swarms, *IEEE Transactions on Automatic Control*, Vol. 48, No. 10, pp. 1848–1854.

[105] Tanner, H. G., Pappas, G. J., and Kumar, V. (2004). Leader-to-formation stability, *IEEE Transactions on Robotics and Automation*, Vol. 20, No. 3, pp. 443–455.

[106] Lohmiller, W., and Slotine, J.-J. E. (1998). On contraction analysis for nonlinear systems, *Automatica*, Vol. 34, No. 6, pp. 683–696.

[107] Boyd, S., El Ghaoui, L., Feron, E., and Balakrishnan, V. (1994). *Linear Matrix Inequalities in System and Control Theory* (SIAM).

[108] Gahinet, P., Nemirovski, A., Laub, A. J., and Chilali, M. (2004). *LMI Control Toolbox- For Use with MATLAB, User's Guide, Version 1.0.9* (The MathWorks, Inc).

[109] VanAntwerp, J. G., and Braatz, R. D. (2000). A tutorial on linear and bilinear matrix inequalities, *Journal of Process Control*, Vol. 10, No. 4, pp. 363–385.

[110] Ren, W. (2007). Consensus strategies for cooperative control of vehicle formations, *IET Control Theory Appl.*, Vol. 1, No. 2, pp. 505–512.

[111] Ren, W., and Beard, R. (2004). Trajectory tracking for unmanned air vehicles with velocity and heading rate, *IEEE Transactions on Control Systems Technology*, Vol. 12, No. 5, pp. 706–716.

[112] Desoer, C. A., and Vidyasagar, M. (1975). *Feedback Systems: Input-Output Properties* (Academic Press).

[113] Shanmugavel, M., Tsourdos, A., White, B. A., and ŻMultiple UAVs, *ASME Journal of Dynamic Systems, Measurement and Control*, Vol. 129, No. 5, pp. 620–632.

[114] Léchevin, N., Rabbath, C. A., and Sicard, P. (2006). Stable Morphing of Unicycle Formations in Translational Motion, in *Proceeding of the American Control Conference*, Minneapolis, MN, pp. 4231–4236.

[115] Léchevin, N., Rabbath, C. A., and Sicard, P. (2006). Trajectory Tracking of Leader-Follower Formations Characterized by Constant Line-of-Sight Angles, *Automatica*, Vol. 42, December, pp. 2131–2141.

[116] Van der Schaft, A. J. (2000). *L2-Gain and Passivity Techniques in Nonlinear Control, LNCIS, Vol. 218* (Springer-Verlag).

[117] Ortega, R., Loria, A., Nicklasson, P. J., and Sira-Ramirez, H. (1998). *Passivity-Based Control of Euler-Lagrange Systems* (Springer-Verlag).

[118] Wen, J. T. (1988). Time Domain and Frequency Domain Conditions for Strict Positive Realness, *IEEE Transactions on Automatic Control*, Vol. 33, No. 10, pp. 988–992.

[119] Ljung, L. (2005). *System Identification Toolbox For Use with Matlab, User's Guide, Version 6* (The MathWorks, Inc).

[120] Shanmugavel, M. (2007). *Path Planning of Multiple Autonomous Vehicles*, PhD Thesis, Defence College of Management and Technology, Cranfield University, UK.

[121] Farouki, R. T. and Sakkalis, T. (1990). Pythagorean Hodographs, *IBM Journal of Research and Development*, Vol. 34, No. 5, pp. 736–752.
[122] Farouki, R. T. and Sakkalis, T. (!994). Pythagorean hodograph space curves, *Advances in Computational Mathematics*, Vol. 2, pp. 41–66.
[123] Léchevin, N., Rabbath, C. A., Tsourdos, A., and White, B. A. (2008). An Integrated Decision, Control and Fault Detection Scheme for Cooperating Unmanned Vehicle Formations, in *Proceedings of American Control Conference*, Seattle, WA, pp. 1997–2002.
[124] *Scilab 5.0* `http://www.scilab.org/`.
[125] Giulietti, F., Pollini, L., and Innocenti, M. (2000). Autonomous Formation Flight, *IEEE Control Systems Magazine*, Vol. 20, December, pp. 34–44.
[126] Pollini, L., Giulietti, F., and Innocenti, M. (2002). Robustness to Communication Failures within Formation Flight, in *Proceedings of the American Control Conference*, Anchorage, AK, pp. 2860–2866.
[127] Innocenti, M., and Pollini, L. (2004). Management of Communication Failures in Formation Flight, *AIAA Journal of Aerospace Computing, Information, and Communication*, Vol. 1, No. 1, pp. 19–35.
[128] Mehra, R. K., Boskovic, J. D. and Li, S.-M (2000). Autonomous formation flying of multiple UCAVs under communication failure, in *IEEE Position Location and Navigation Symposium*, pp. 371–378.
[129] Valenti, M., Bethke, B., Fiore, G., How, J., and Feron, E. (2006). Indoor multi-vehicle flight testbed for fault detection, isolation, and recovery, in *Proc. of AIAA Guidance, Navigation, and Control Conference and Exhibit*, Keystone, Colorado, AIAA-2006-6200.
[130] Meskin, N. and KHorasani, K. (2009). Actuator Fault Detection and Isolation for a Network of Unmanned Vehicles, *IEEE Transactions on Automatic Control*, Vol. 54, No. 4, pp. 835–840.
[131] Daigle, M. J., Koutsoukos, X. D., and Biswas, G. (2007). Distributed diagnosis in formations of mobile robots, *IEEE Transactions on Robotics*, Vol. 23, No. 2, pp. 353–369.
[132] Ferrari, R. M., Parisini, T., and Polycarpou, M. M. (2009). Distributed fault diagnosis with overlapping decompositions: an adaptive approximation approach, *IEEE Transactions on Automatic Control*, Vol. 54, No. 4, pp. 794–799.
[133] Chung, W. H. and Speyer, J. L. (1998). A Decentralized Fault Detection Filter, in *Proceedings of the American Control Conference*, pp. 2017–2021.
[134] Shankar, S., Darbha, S., and Datta, A. (2002). Design of a decentralized detection filter for a large collection of interacting LTI systems, *Math. Problems Eng.*, Vol. 8, No. 3, pp. 233–248.
[135] Liang, F., Wang, J. L., and Yang, G.-H. (2002). Reliable Robust Flight Tracking Control: An LMI Approach, *IEEE Transactions on Control Systems Technology*, Vol. 1, No. 1, pp. 76–89.
[136] Emami-Naeini, A., Akhter, M. M., and Rock, S. M. (1988). Effect of Model Uncertainty on Failure Detection: the Threshold Selector, *IEEE Transactions on Automatic Control*, Vol. 33, No. 12, pp. 1106–1115.
[137] Tuan, H. D., Apkarian, P., and Nguyen, T. Q. (2005). Robust Filtering for

Uncertain Nonlinearly Parameterized Plants, *IEEE Transactions on Signal Processing*, Vol. 51, No. 7, pp. 1806–1815.

[138] Léchevin, N., and Rabbath, C. A. (2007). Robust decentralized fault detection in leader-to-follower formations of uncertain, linearly parameterized systems, *AIAA Journal of Guidance, Control and Dynamics*, Vol. 30, No. 5, pp. 1528–1535.

[139] Åström, K. J., and Wittenmark, B. (1984). *Computer Controlled Systems: Theory and Design* (Prentice-Hall).

[140] Tuan, H. D., Apkarian, P., and Nguyen, T. Q. (2001). Robust and reduced-order filtering: new LMI-based characterizations and methods, *IEEE Transactions on Signal Processing*, Vol. 49, No. 12, pp. 2975–2984.

[141] Wang, Z. and Willett, P. K. (2001). All-purpose and plug-in power-law detectors for transient signals, *IEEE Transactions on Signal Processing*, Vol. 49, No. 11, pp. 2454–2466.

[142] Poor, H. V. (1994). *An Introduction to Signal Detection and Estimation* (Springer).

[143] Earon, E. (2006). Almost-lighter-than-air vehicle fleet simulation - Technical Report V. 1.1 (Quanser Inc.).

[144] Balas, G., Chiang, R., Packard, A., and Safonov, M. (2009). *Robust Control Toolbox 3 - User's Guide* (The MathWorks, Inc).

[145] Juang, J.-G., and Cheng, K.-C. (2006). Application of Neural Networks to Disturbances Encountered Landing Control, *IEEE Transactions on Intelligent Transportation Systems*, Vol. 7, No. 4, pp. 582–588.

[146] Meskin, N., Jiang, T., Sobhani, E., Khorasani, K., and Rabath, C. A. (2007). A nonlinear geometric fault detection and isolation approach for almost-lighter-than-air-vehicles, in *Proc. IEEE Multi-Conference on Systems and Control*, Suntec City, Singapore, pp. 1073–1078.

[147] Chen, T. and Francis, B. (1995) *Optimal Sampled-Data Control Systems* (Springer-Verlag).

[148] Goodwin, G. C. and Middleton, R. H. (1990). *Digital Control and Estimation - A Unified Approach* (Prentice-Hall).

[149] Lemmon, M. and Ling, Q. (2004). Control Systems Performance under Dynamic Quantization: the scalar case, in *Proc. 43rd IEEE Conference on Decision and Control*, Atlantis, Bahamas, pp. 1884–1888.

[150] Rabbath, C. A., Léchevin, N. and Hori, N. (2004). Optimal Dual-Rate Digital Redesign with Application to Missile Control, *Journal of Guidance, Control and Dynamics*, Vol. 27, No. 6, pp. 1083–1087.

[151] Ho, Y. C. and Pepyne, D. L. (2002). Simple Explanation of the No-Free-Lunch Theorem and Its Implications, *Journal of Optimization Theory and Applications*, Vol. 115, No. 3, pp. 549–570.

[152] Krokhmal, P., Murphey, R., Pardalos, P., and Uryasev, S. (2004). *Use of Conditional Value-at-Risk in Stochastic Programs with Poorly Defined Distribustions*, ch. in *Recent Developments in Cooperative Control and Optimization*, Butenko, S., Murphey, R., and Pardalos, P. (eds.) (Kluwer Academic Publisher).

[153] Shima, T., Ramussen, S. J., and Sparks, A. G. (2005). UAV Cooperative

Multiple Task Assignments Using Genetic Algorithm, in *Proceedings of the American Control Conference*, Portland, OR, pp. 2989–2994 .

[154] Kim, Y., Gu, D., and Postlethwaite, I. (2007). *Real-Time Optimal Time-Critical Target Assignment of UAVs*, ch. in *Advances in Cooperative Control and Optimization*, Hirsh, M. J., Pardalos, P. M., Murphey, R., and Grundel, D. (eds.) (Springer).

[155] Léchevin, N., Rabbath, C. A., and Lauzon, M. (2009). *A Distributed Network Enabled Weapon-Target Assignment for Combat Formations*, ch. in *Optimization and Cooperative Control Strategies*, Hirsch, M. J., Commander, C., Pardalos, P. M., and Murphey, R. (eds.) (Springer).

[156] Rubinstein, R. Y. and Kroese, D. P. (2004). *The Cross-Entropy Method - A unified Approach to Combinatorial Optimization, Monte-Carlo Simulation and Machine* (Springer).

[157] *Urban Operations* `http://rdl.train.army.mil/soldierPortal/atia/adlsc/view/public/11645-1/fm/3-06/toc.htm\#toc`.

[158] Léchevin, N., Rabbath, C. A. and Lauzon, M. (2009). A Decision Policy for the Routing and Munitions Management of Multiformations of Unmanned Combat Vehicles in Adversarial Urban Environments, *IEEE Transactions on Control Systems Technology*, Vol. 17, No. 3, pp. 505-519.

[159] Cruz, J. B. Jr., Simaan, M. A., Gacic, A., Jiang, H., Letellier, B., Li, M., and Liu, Y. (2001). Game-Theoretic Modeling and Control of a Military Air Operation, *IEEE Trans. Aerosp. Electron. Syst.*, Vol. 37, No. 4, pp. 1393–1403.

[160] Cruz, J. B. Jr., Simaan, M. A., Gacic, A., and Liu, Y. (2002). Moving Horizon Nash Strategies for a Military Air Operation, *IEEE Trans. Aerosp. Electron. Syst.*, Vol. 38, No. 3, pp. 989–997.

[161] Galati, D. G. and Simaan, M. A. (2003). Effectiveness of the Nash Strategies in Competitive Multi-Team Target Assignment Problems, in *Proc. IEEE Conf. Decision and Control*, Paradise Island, Bahamas, pp126–134.

[162] Ghose, D., Krichman, M., Speyer, J. L., and Shamma, J. (2002). Modeling and Analysis of Air Campaign Resource Allocation: A Spatio-Temporal Decomposition Approach, *IEEE Trans. Syst., Man, Cybern. - Part A: Systems and Humans*, Vol. 32, No. 3, pp. 403–418.

[163] Basar, T. and Bernhard, P. (1995). *H^{∞}-Optimal Control and Related Minimax Design Problems* (Birkhäuser).

[164] McEneaney, W. M., Fitzpatrick, B. G., and Lauko, I. G. (2004). Stochastic Game Approach to Air Operations, *IEEE Trans. Aerosp. Electron. Syst.*, Vol. 40, No. 4, pp. 1191–1216.

[165] McEneaney, W. M. and Singh, R. (2004). Unmanned Vehicle Operations under Imperfect information in an Adversarial environment, in *Proc. AIAA 3rd Unmanned Unlimited Technical Conference, Workshop and Exhibit*, Chicago, Illinois, AIAA-2004-6411.

[166] McEneaney, W. and Singh, R. (2005). Deception in Autonomous Vehicle Decision Making in an Adversarial Environment, in *Proc. AIAA Guidance, Navigation, and Control Conference and Exhibit*, San Francisco, California, AIAA-2005-6152.

[167] Beard, R. W., McLain, T. W., Goodrich, M. A., and Anderson, E. P. (2001). Coordinated target assignment and intercept for unmanned air vehicles, *IEEE Trans. Robot. Automat.*, Vol. 9, No. 6, pp. 777–790.

[168] Richards, A. and How, J. P. (2002). Aircraft Trajectory Planning with Collision Avoidance Using Mixed Integer Linear Programming, in *Proc. Amer. Control Conf*, Anchorage, Alaska, pp. 1936–1941.

[169] Dogan, A. (2003). Probabilistic Path Planning for UAVs, in *Proc. 2nd AIAA Unmanned Unlimited Systems, Technologies, and Operations*, San Diego, California, AIAA-2003-6552.

[170] Chasparis, G. C. and Shamma, J. S. (2005). Linear-Programming-Based Multi-Vehicle Path Planning with Adversaries, in *Proc. Amer. Control Conference*, Portland, OR, pp. 1072–1077.

[171] Flint, M., Fernandez-Gaucherand, E., and Polycarpou, M. (2003). Stochastic Modeling of a Cooperative Autonomous UAV Search Problem, *Military Operations Research Journal*, Vol. 8, No. 4, pp. 13–32.

[172] Bertsimas, D. J. and Van Ryzin, G. J. (1993). Stochastic and Dynamic Vehicle Routing in the Euclidean Plane with Multiple Capacited Vehicles, *Operations Research*, Vol. 41, No. 1, 1993, pp. 60–76.

[173] Frazzoli, E. and Bullo, F. (2004). Decentralized Algorithms for Vehicle Routing in a Stochastic Time-Varying Environment, in *Proceedings of the IEEE Conf. Decision and Control*, Atlantis, Paradise Island, Bahamas, pp. 3357–3363.

[174] Olfati-Saber, R. (2006). Flocking for Multi-Agent Dynamic System: Algorithms and Theory, *IEEE Trans. Autom. Control*, Vol. 51, No. 3, pp. 401–420.

[175] Cortes, J., Martinez, S., and Bullo, F. (2006). Robust Rendezvous for Mobile Autonomous Agents via Proximity Graphs in Arbitrary Dimensions, *IEEE Trans. Autom. Control*, Vol. 51, No. 8, pp. 1289–1298.

[176] Toth, P., and Vigo, D. (eds.) (2002). *The Vehicle Routing Problem* (SIAM).

[177] Godsile, C. and Royle, G. (2001). *Algebraic Graph Theory* (Springer-Verlag).

[178] Nilim, A. and El Ghaoui, L. (2005). Robust Control of Markov Decision Processes with Uncertain Transition Matrices, *Operations Research*, Vol. 53, No. 5, pp. 780–798.

[179] Dupuy, T.N. (1995). *Attrition: Forecast Battle Casualties and Equipment Losses in Modern War* (Nova).

[180] Basar, T. and Olsder, G. J. (1999). *Dynamic Noncooperative Game Theory* (SIAM).

[181] Mahler, R. (2004). *Objective functions for Bayesian ontrol-Theoretic Sensor Management, II: MHC-like Approximation*, ch. in *Recent Developments in Cooperative Control and Optimization*, Butenko, S., Murphey, R., and Pardalos, P. (eds) (Kluwer Academic).

[182] Bertsekas, D., and Tsitsiklis, J. (1996). *Neuro-Dynamic Programming* (Athena Scientific).

[183] Olfati-Saber, R., Fax, J. A., and Murray, R. (2007). Consensus and Cooper-

ation in Networked Multi-Agent Systems, *Proceedings of the IEEE*, Vol. 95, No. 1, pp. 215–233.

[184] Ren, W., Beard, R. W., and Atkins, E. M. (2007). Information Consensus in Multivehicle Cooperative Control, *IEEE Control Systems Magazine*, Vol. 27, No. 2, pp. 71–82.

[185] Moallemi, C. C., and Van Roy, B. (2006). Consensus Propagation, *IEEE Transactions on Information Theory*, Vol. 52, No. 11, pp. 4753–4766.

[186] Mosk-Aoyama, D., and Shah, D. (2006). Computing Separable Functions via Gossip, in *Proceedings of the 25th annual ACM Symposium on Principles of Distributed Computing*, Denver, Colorado, pp. 113–122.

[187] Cortes, J. (2006). Finite-Time Convergent Gradient Flows with Applications to Network Consensus, *Automatica*, Vol. 42, No. 11, pp. 1993–2000.

[188] Sundaram, S. and Hadjicostis, C. N. (2008). Distributed Functional Calculation and Consensus using Linear Iterative Strategies, *IEEE Journal on Selected Areas in Communications*, Vol. 26, No. 4, pp. 650–660.

[189] Léchevin, N., Rabbath, C. A., and Zhang, Y. (2009). Information Broadcasting Algorithm for Finite-Time Reaching-at-Risk Consensus with Application to Weapon-Target Assignment, in *Proceedings of the American Control Conference*, St. Louis, MO.

[190] Fleming, W. H., and McEneaney, W. M. (2001). Robust Limits of Risk Sensitive Nonlinear Filters, *Math. Control, Signals and Systems*, Vol. 14, pp. 109–142.

[191] Matsumoto, M. and Nishimura, T. (1998). Mersenne Twister: A 623-dimensionally equidistributed uniform pseudorandom number generator, *ACM Trans. on Modeling and Computer Simulation*, Vol. 8, No. 1, pp. 3–30.

[192] Robert, C. P. and Casella, G. (1999). *Monte Carlo statistical methods* (Springer).

[193] Bertsekas, D. (1982). Distributed Dynamic Programming, *IEEE Transactions on Automatic Control*, Vol. 27, No. 3, pp. 610–616.

[194] Jalali, A. and Ferguson, M. J. (1992). On Distributed Dynamic Programming, *IEEE Transactions on Automatic Control*, Vol. 37, No. 5, pp. 685–689.

[195] Bertossi, A. A. and Mei, A. (2000). Constant Time Dynamic Programming on Directed Reconfigurable Networks, *IEEE Transactions on Parallel and Distributed Systems*, Vol. 11, No. 6, pp. 529–536.

[196] Canto, S. D., de Madrid, A. P., and Bencomo, S. D. (2005). Parallel Dynamic Programming on Clusters of Workstations, *IEEE Transactions on Parallel and Distributed Systems*, Vol. 16, No. 9, pp. 785–798.

[197] Liu, W. and Schmidt, B. (2006). Parallel Patterns-Based Systems for Computational Biology: A Case Study, *IEEE Transactions on Parallel and Distributed Systems*, Vol. 17, No. 8, pp. 750–763.

[198] *RT-LAB User's Guide* (Opal-RT Technologies, Inc.).

[199] *Simulink 7 User's Guide* (The MathWorks, Inc.).

[200] *QuaRC 1.0 Installation Guide* (Quanser, Inc.).

[201] *RedHawk Linux* (Concurrent Computer Corporation) `http://www.ccur.com`.

Index